What is Safe?

The Risks of Living in a Nuclear Age

What is Safe?
The Risks of Living in a Nuclear Age

David R. Williams
Department of Chemistry, University of Wales, Cardiff, UK

THE ROYAL
SOCIETY OF
CHEMISTRY
Information
Services

ISBN 0-85404-569-4

A catalogue record for this book is available from the British Library.

Published by The Royal Society of Chemistry,
Thomas Graham House, Science Park, Milton Road, Cambridge CB4 4WF, UK

For further information see our web site at www.rsc.org

Typeset by Paston Press Ltd, Loddon, Norfolk
Printed by Redwood Books Ltd, Trowbridge, Wiltshire

Preface

This book arises from three decades of experience teaching science to chemists, dentists, medics, pharmacists, and to lay-persons and from speaking in a series of radio and TV broadcasts.

It is axiomatic that, frequently, in explaining the solution to one problem, another is created. In this respect, the objective of answering 'What is safe?' cannot be completed in one book or even in a series of books but, rather, we aim to explain the fundamental language of the problems faced in modern-day life and in future decisions that need to be taken such that these problems can be better described and thus more easily tackled.

A significant proportion of the discussions and of the data recorded in this book are from the cancer research and from the nuclear and radwaste fields. Not only are these the areas in which the majority of hard, numerate facts have been recorded and published in well-refereed journals over the last few decades, but also these are topics which have come from obscurity to maturity within the living memories of many of us. We are now in the sixth decade of nuclear power. Similarly, extremely powerful computers now affect the decision-making and statistic gathering activities of everyday lives, and yet in 1961 the computer power of all the computers in the World was less than that of the modern cheap wristwatch now worn by billions. Fortunately, the principles of the technology underlying the data treatment are transferable to new areas. For example, the recent suggestion that there may well be clusters of CJD traceable to water contamination will be subject to the same critical assessment procedures as were used in examining the postulated cancer 'clusters' near nuclear sites at Dounreay and Sellafield. The technology is indeed extremely transferable.

In mentioning food and diet, it is significant that in the days of Descartes, 'I think therefore I am', the mind was nurtured and the body was left to fend for itself, whereas nowadays, the almost religious zeal with which some search for the perfect diet to give perfect

v

health of body and soul has produced a profound change of dietetic emphasis.

Many would accept that the three greatest threats to a stable, modern society are urban terrorism and/or fundamentalism, the destruction of our environment, and nutritional imbalance arising from the increasing population. There is a steadily widening gap and a polarization between the 'very rich' and the 'very poor'.

By and large, the 'haves' and the 'have nots' fall into three broad classes – those who are getting *on*, those who are just getting *by*, and those who are getting *nowhere*. All lives are affected by decisions concerning morbidity and mortality and we have a right to be kept informed of the myriad of factors which affect our lifestyles.

Unfortunately, there is often a language and perception barrier to the transference of information and data from the specialist to the lay-person. In terms of perception we have to decide how much reliability and credence we must place upon statements from chartered scientists, from advertisers, from members of the public, or from journalists. We would do well to remember the statement made by scientists announcing the results of the analysis of the radiocarbon dating of the Turin shroud in 1989, 'If we accept a scientific result, we must exercise a critical notion of the probabilities involved. If we demand absolute certainty, we shall have to rely on faith.'

This book attempts to tackle these topics both by describing terms in widespread use by so-called experts and also by providing overviews of various items. Indeed, like many of the emotive subjects, we echo the words of James Thurber, 'So much has already been written about everything that you cannot find out anything about it.'

I am most grateful to the scientists and engineers of NIREX who have offered much data and advice (they are at the cutting edge of quantitative statistic gathering and researching in this area), to the group of specialists who published a booklet through the British Medical Association entitled *Living with Risks* (they are at the forefront of communicating risks and their perception to patients), and to my colleagues, students, family and audiences who have listened and asked questions which have often produced greater clarity in my mind.

Thank you all very much indeed.

David R. Williams

Contents

	Abbreviations	viii
1	Introduction	1
2	Daily Life	4
3	Childhood	6
4	Transport Risks	7
5	Homes	10
6	Recreation	14
7	Healthcare	15
8	Diet	18
9	Medical	20
10	Energy	23
11	Perception	29
12	Risk, Perception and Social Constructions	38
13	Risk Management	45
14	Risks at Work	48
15	Risks from Chemicals	50
16	General Principles of Risk	54
17	Defining 'Safe' and 'Safe Enough'	59
18	Statistics – Dealing with Large Numbers	61
19	Radioactive Waste Disposal	70
20	Modelling Doses and Risks	88
21	Marshalling our Facts	97
22	Perception of Radwaste Disposal	102
23	Concluding Remarks	115
	Appendix	119
	Further Reading	135
	Subject Index	137

Abbreviations

ACSNI	Advisory Committee on the Safety of Nuclear Installations
AEA	Atomic Energy Authority
AGR	Advanced Gas-cooled Reactor
ALARA	As Low As Reasonably Achievable
ALARP	As Low As Reasonably Practicable
BMA	British Medical Association
CEGB	Central Electricity Generating Board
COMARE	Committee on the Medical Aspects of Radiation in the Environment
HACCP	Hazard Analysis Critical Control Point
HLW	High Level Waste
HSE	Health and Safety Executive
ICRP	International Commission on Radiological Protection
ILW	Intermediate Level Waste
LLW	Low Level Waste
NEL	No adverse Effects Level
NIMBY	Not In My Back Yard
NIREX	Nuclear Industry Radioactive Waste Executive
NRPB	National Radiological Protection Board
OECD	Organization for Economic Cooperation and Development
OPCS	Office of Population Censuses and Surveys
RWMAC	Radioactive Waste Management Advisory Committee
SDU	Safety Degree Unit
SLILW	Short-lived Intermediate Level Waste
SMR	Standardized Mortality Ratio
TOR	Tolerability of Risk
TLV	Threshold Limit Value

What is Safe?

The Risks of Living in a Nuclear Age

1 Introduction

'Safe' is associated with many considerations, most of which can be summarized in terms of four-letter words such as feel and fear (as in perception), risk, dose, work, play, life, dead (as in death), and cost – even safe itself is a short word that disguises the exceedingly complex manifestations that go to make up the definition of safe.

Dictionary definitions of 'safe' include "offering security or protection from harm", "free from danger", US, a slang word for condom (Collins English Dictionary) and "free from hurt or damage", "having escaped some real or apprehended danger" (Shorter Oxford English Dictionary).

Human nature, personality and perception are very much involved with one's accepted definition of the word 'safe'. 'Safe' is often taken as the reciprocal of a magnitude of a risk. Humans tend to overestimate a risk when referring to rather unusual, rare or exceptional incidents but tend to underestimate a risk markedly when common, everyday events are involved. The triangular diagram (Figure 1.1) shows how safety is a compromise and is inter-disciplinary between the three-cornered considerations of risk, benefit and cost.

Difficulties are immediately encountered when one wishes to put a scale of numbers on to these diagrams, such as the origin implying a state of zero risk, *i.e.* absolute safety, which does not exist. Similarly, different humans would have different impressions of the units and relative

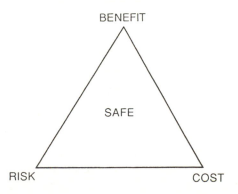

Figure 1.1 *Risk, benefit and cost enveloping a safe situation*

importances of these risks and so we soon enter into subjective view-points and human values.

The traditional approach to attempting to quantify safety is to collect data concerning risks. Here one soon encounters difficulty in deciding whether these data should be in terms of deaths, length of life, or quality of life. The modern viewpoint seems to be that risk is best quantified in terms of a combination of data concerning both the quality and the length of the subject's life. This underlines the fact that we would far prefer to *live* our lives rather than *merely to exist*. There are many public figures who, when questioned about their lifestyle and whether it was safe, have publicly declared that they give zero consideration to trying to extend their lives, but rather they dedicate it to using their lifetime to maximum effect. This all suggests that they are disregarding the safest lifestyle for one which is far more risky and endangers their safety, one that either gives them great gratification from its stimulations or does a great deal of good for others through their charitable actions. These are individual and personal choices.

Most of us never think of using the word 'risk' until we hit a dilemma concerning the safety of our lives decided by other groups of people; for example, if they decide to build a facility in our neighbourhood or expose us to some threat that will possibly change our lifestyle. It is at such times that we clutch for the word 'safe' and begin to worry whether the basis upon which the decision has been reached is scientifically sound, whether the communication of the reasons underpinning that decision has been sufficiently clear, and whether, if we had played our full part in the decision process, we would have recommended differently to the body that has made the decision to impose an involuntary risk upon us.

It is about such topics that this book is directed. It is written from the viewpoint of a chemist and a technologist and, perhaps, such profes-sionals ought to accept a greater responsibility for providing data to decision-making bodies in forms which are more helpful. After all, we produce in excess of 1000 new chemicals per year with a view to their marketing. The text lists many illustrations that can be used when talking to lay-persons. Several are taken from the British Medical Association book entitled *The BMA Guide to Living with Risk* and the BMA has pointed out that, in the eyes of the lay public, the healthy, enjoyable, stimulating scientific debate in which we scientists participate is seen as confusion amongst professionals. This generates fear and uncertainty in the minds of the public.

This leads us to the subject of familiarity. Because they are not familiar with the workings of the minds of scientists and technologists they see things very differently to our perception. The same familiarity with everyday risks permits one to enjoy drinking and swallowing and smoking carcinogenic, addictive, poisonous materials as part of leisure activities with little concern for their effects.

The universal observation that experts are distrusted worldwide is dwarfed by the fact that their statistics are distrusted even more so! Indeed, there seems to be only one certain indisputable fact – there is nothing certain about life except death!

In numerical terms, 'Risk = magnitude of a consequence × the probability of its occurrence', but one ought not immediately turn to quantifying risks because that is often easier than a more fundamental consideration of the significance of any risk data gathering activity that is being contemplated. The perception and emotional aspects of what is being investigated have a great bearing on the overall risk envisaged.

Even the concept of 'acceptable risk' has ill-defined margins. There is all the difference in the world between something being acceptable and a risk which has to be accepted. Although everyone agrees that premature death is not to be welcomed, there is great disagreement, dependent upon circumstances, concerning the definition of the word 'premature'. To an elderly person, any death before tomorrow seems to be premature. We have quoted the dictionary definitions of 'safe'. Another term that is widely used in this article is that of 'risk' which is defined in the dictionary as "the possibility of incurring misfortune or loss; hazard" (Collins English Dictionary), or "hazard, danger or exposure to mischance or peril" (Shorter Oxford English Dictionary).

The UK Health and Safety approach to Tolerability of Risk (TOR) has a more formal definition and is referred to later (Section 16). The origin is believed to be from the Greek word 'rhiza' which was in widespread use when discussing the hazards of ships sailing too close to the cliffs. Pharmacists remind us that Hypocrates in the fourth century stated that sharp objects are more likely to cause injury because forces were concentrated at a point. There is a parallel in modern times with nuclear and other forms of power which concentrate energy within a very small volume.

Although scientific data have long been collected in an orderly and well-planned fashion, it is only with the development of probability theory by Pascal in the mid-seventeenth century that the likelihood of events occurring could be calculated. Although it is claimed that this was the turning point when scientists analysed and made future predictions rather than relying upon soothsayers, it is perhaps sad to note the resurgence of interest in astrological predictions and the like during the last decade or so.

From data collection, causal links were soon identified between tobacco, snuff and cancer of the nose lining, between sunlight and skin cancer, between London smogs and respiratory disease, and also the historically important link between contaminated water and cholera identified by Farr, Snow, Chadwick *et al.*

Chadwick and Farr's research led to a listing of causes of death in 1841. Nowadays, the Office of Population Censuses and Surveys uses the standardized mortality ratio (SMR) which describes the number of deaths registered in a particular period of time as a percentage of those expected in a selected year at the age and sex mortality rates operated in the year of interest. For example, infant mortality has dropped over the last 150 years from 148 per 1000 to 9 per 1000 (from 1 in 7 to 1 in 106). In fact, the main improvement in life expectancy which has occurred during this century has been due to this great reduction in the death of neonates. Sadly, two-thirds of deaths amongst males around about the age of twenty now occur because of violence and/or accidents. After the age of forty, circulatory disease takes over and, overall, one in four deaths occur through cancer.

Technology has not only offered a cornucopia of new benefits but also new risks. Technology now influences our everyday lives through decisions concerning events and new innovations. Such are the stringent checks upon the introduction of new technology that modern-day life is now considerably safer than it was half a century ago. The complexity, however, of the prevailing technology means that life is no longer as simple as it used to be, but it is certainly safer.

2   Daily Life

The term 'safe' is related to risks and events that occur in normal, everyday life supplemented with incidents which are unexpected, unusual, or fairly random in their occurrence.

Even the pattern of normal, everyday life will vary from person to person, dependent upon their age, position in life, the country in which they dwell, their self-selected lifestyle and habits, *etc.*

However, there are certain common features to this background pattern of daily life to which we are all committed. For example, we all have to be born (which involved a certain amount of risk); we are all exposed to a range of risks (although their occurrence may well be determined by our own decisions), for example, travelling, smoking tobacco, infection, *etc.*; and we all eventually succumb to death from the residual of any conditions which have not been successfully tackled during early years.

In fact, death is 100% certain and inevitable and, in absolute terms, is not preventable when it occurs at the end of a long life. In relative terms, the expected life span at birth is expanding, such that we live an average

30 years longer than our Victorian forebears because other threats to life are being reduced through preventative medicine, therapy, information and education, and risk management. The commonest causes of death in the Western World nowadays are cardiovascular disease and cancer. These are the residual legatees after all other threats to middle age and to youth have been reduced.

This section describes certain high-rise features of elevated risk of death occurring throughout a 'normal' lifestyle and, also, discusses the *quality of life* throughout a lifetime which is possibly of more importance to the population than the actual *cause of death* at the end of a long and satisfactory lifespan.

'Being safe' is about decisions that affect the length of our lives until death and also the quality of the life between birth and death; these areas are known as mortality and morbidity studies, respectively. Decisions which enhance the quality of our life and happiness usually involve certain risks. Sometimes reducing the risk of an event will improve the happiness (for example, removing a cataract in order to enhance a person's chances of safely crossing the road) and sometimes the decision to improve the quality or the excitement of life will significantly increase the risk (for example, deciding to exceed the speed limit whilst driving a motor vehicle).

Causes of death have been recorded on death certificates since the year 1841, and since standard mortality ratios were introduced into the statistics of dying, various theories linking lifestyle during normal everyday life and cause of death have been used to justify changes in procedures; this is reflected in the increasing life expectancy for males and females shown in the diagram (Figure 2.1). What these figures do not reflect is the enjoyment of the quality of life during the many decades before death. Also, they do not reflect the number of early deaths that can be shown to have a causal link with personal decisions taken by the deceased or by collective decisions imposed upon a community by, for example, a governing authority granting permission to construct nearby.

It is axiomatic that we are challenged by fewer risks to health nowadays than in any previous generation. Further, we have a far greater control over these risks than at any prior time. Paradoxically, as Wildavsky and Lee have pointed out, 'The richest, longest-lived, best protected, most resourceful civilization, with the highest degree of insight into its own technology, is on the way to becoming the most frightened', and so greater interest is being taken in risk decisions.

The following sections describe some of the main areas of life where persons, or their immediate family, have a considerable degree of control concerning risk decisions which directly affect the quality of life.

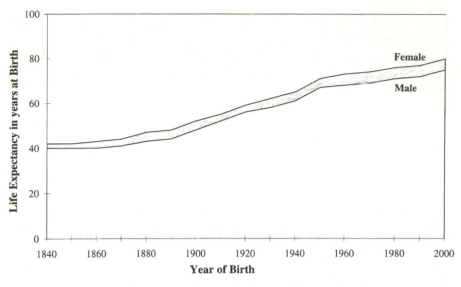

Figure 2.1   *Life expectancy at birth for males and females*
(*Source*: OPCS, 1985 and 1996)

Examples include childbirth and childhood, transport, recreation, home-based risks, diet and medicine. We shall then touch upon subjects where we have far less personal control, such as natural disasters and the general pattern of causes of death for the western industrialized society.

3   Childhood

After parturition, the most prevalent cause of death between 28 days and 12 months is Sudden Infant Death Syndrome (cot death). This occurs for between two and three per 1000 live births. Details concerning the underlying causes are still at the level of hypotheses, several concerning the incidence of upper respiratory tract infections. Moves to reduce these tragic numbers have involved the positioning of the child within the cot and also the introduction of monitoring devices to alert parents.

Between the ages of 1 and 14 years, approximately three British children per day die from accidents. Over and above these 700–1000 fatal injuries per year, there are 120 000 injured child admissions to hospital, approximately 10 000 of whom are permanently disabled. Clearly, their quality of life is severely reduced because of these incidents. Although the number of accidents arising from organized club and educational and recreational activities, such as canoeing and mountaineering, *etc.*, per 1000 children being involved in such schemes has levelled off, accidents to children in the home continue to be a major cause for concern, there being up to 1 000 000 injuries (thankfully, many of them minor) per year.

Outside the home an even greater number, approximately 1.2 million per year, are injured. Road traffic accidents are the cause of death for the order of 800 children per year, some 15% of which are linked to cycling, but for two-thirds the child was a pedestrian.

Some of the accidents to children both inside and outside homes are related to social class factors. This concept is carried through to the environmental damage inflicted upon children. Those in the lower social classes have more limited sporting and playing facilities, and safety provisions. Improved safety for children goes hand-in-hand with improving living conditions and the provision of safe leisure activities for all.

4   Transport Risks

Although there is an obvious qualitative link between the distance and the number of persons travelling or the amount of goods being transported, to lowering of life quality or to death, quantitative statistics must correlate data with mode of transport (car, plane, cycle, *etc.*), and with exposure (kilometres, hours). There are often important secondary links; for example, the stress of being a driver gridlocked on the M25 London orbital motorway has a primary collision risk of approximately zero but the chances of an accident once released from the jam and of stress related health conditions later are considerably raised.

Statistically, high-risk factors concerning road traffic deaths include being in a rural area as distinct to urban (speeds are higher and emergency assistance is more distant, *e.g.* even a small population such as Wales has 10 000 road traffic accidents per year), travelling at night, which is roughly three times as dangerous as day-time travel (influence of alcohol and poor visibility on unfamiliar roads), and weekend travel which accounts for 50% of all road fatalities (unfamiliar drivers distracted by the relish of the event to which they drive). Clearly, the time travelled is also a major factor.

When one examines the statistics of road traffic accidents there are two important principles revealed which apply to all estimations of risk throughout this document:

(1) Media Reporting Bias
   The widespread misconception which arises from headline and front-page colour photograph reporting is of horrendous, multiple vehicle pile-ups and burn-outs on high speed roads such as motor-

ways. In fact, such occurrences happen exceedingly rarely and contribute only a very small number of death statistics to the data.

(2) Statistic Gathering Difficulties

It is extremely difficult to acquire reliable statistics concerning road accidents, deaths and injuries. Although the number of deaths can be counted, it is difficult to express them in terms of the total number of accidents occurring on the roads, since most of those not involving injury do not get logged. Secondly, the victims of an accident are often treated in hospitals or through GP's surgeries in widely-separated areas employing different means of reporting. Further, pedestrians involved in road traffic accidents are counted differently rather than in terms of the distance that they have travelled.

The BMA has studied road traffic accident data for the 90 years since road crash statistics were first collected in the UK; in 1909, there were 1070 reported fatal accidents which accounted for one crash per year for every 90 vehicles on the road! That is, 106 fatal accidents for every 10 000 vehicles. Nowadays, it is the order of one fatal accident for every 4700 vehicles. These data reflect another important principle concerning the statistical viewpoint – a person crossing a road in 1909 had a much lower average probability of being involved in an accident with a motor vehicle than crossing the same road nowadays. However, on the rare occasion that a motor vehicle used that road in the early part of this century, there was a far higher risk of the accident occurring and of it being fatal.

Clearly, there has been an unbelievable increase in the number of motor vehicles on our roads (there are believed to be some 25 million vehicles, including 400 000 lorries, of which some 5 million are either unroadworthy, and/or unlicensed, and/or uninsured) and the reported road traffic deaths for 1996 were 3600. This accident and death rate has been markedly reduced by the introduction of motorways and their equivalent, such as autobahns elsewhere, and by the emergency teams being better equipped and faster at reaching the incident by using helicopters, *etc*. In terms of the distance travelled, motor cyclists are 26 times more likely to die from a road traffic accident. In common with most road traffic accidents, increased risk features include slippery surfaces, bad visibility, Saturday and Sunday night travel, raised blood alcohol levels, rush-hour travel, holiday periods, and over-familiarity with the route (most accidents occur within 20 miles of home).

So many important decisions concerning 'safe' are tied in with 'risk' and also another four-letter word, 'cost'. This is illustrated by road crashes in the UK in which in excess of one in every 10 000 people die from such accidents per year (equivalent to 10–15 of the largest sized airliners crashing per year!) and the estimated cost per head of the total population is estimated at £100–200 per person per year. This illustrates the *familiarity principle* that the costs, the risks and the death rates are accepted as everyday occurrences and do not appear to influence pedestrians who quite happily risk crossing in places other than pedestrian crossings, cross on the red light, or motorists who shoot across the stop lights.

Per distance travelled, the approximate risk of travelling by train is one half of that associated with travelling on the roads (figures vary markedly), but for aircraft travel there is a very low-risk rate of one in 2.4×10^9 passenger km (which is eight times the distance to the Sun and back or equal to approximately 60 000 orbits of the World!). This implies that, per kilometre, aircraft passenger travel is 32 times safer than road traffic travel. However, the risks of travelling by road or rail to an unfamiliar airport, the tension of fear of flying, the reduced cabin pressure, and the exposure to radiation from the atmosphere are all voluntary additional risks that have to be taken into account with the statistics for this low-risk style of travelling. Clearly, the most dangerous aspects of air travel are take-off and landing and, even for these stages, it has been calculated that statistically one fatal person accident will occur on average for every 1.5 million flight stages, *i.e.* 3 million take-offs plus landings. This risk figure, which is less than one in a million, will be regarded as 'safe' elsewhere in this article. *The Independent on Sunday*, 24 November 1996, raised the question of whether the Channel Tunnel can be branded as 'safe'. In the UK each year, 3600 people die in road accidents. Ten years ago, 193 people died in the Herald of Free Enterprise sinking, six years ago 180 burned to death on the Scandinavian Star Ferry, two years ago 900 drowned in the Estonia. Accidents in the home account for the loss of 4000 lives in the UK each year. The Channel Tunnel, on the other hand, had a fire accident in which 34 people aboard the train escaped, some 19 requiring hospital treatment. The Channel Tunnel company has banned more than 1000 hazardous substances from passage through its tunnel and is re-examining the carriage of poly(styrene) since the fire of November 1996. Although those on the train had a very frightening experience, it does not compare with the previously mentioned deaths and can only be compared with difficulty to other non-fatal accidents. It does, however, underline the fact that everything in life, even doing nothing, carries an element of risk. A single accident does not necessarily justify a complete re-think of the assumptions upon which the safety criteria have been based, either for the Channel Tunnel or for any other hazard. However, it does require that checks to ensure against these types of risk have been considered.

5   Homes

Homes pose a fascinating challenge both in terms of their perception, location and incumbent risks therein. The North American concept of tying yellow ribbons around trees to welcome home travellers and loved ones to a safe haven, although having an idyllic charm, misrepresents the true science of risk therein. The romantic concept of being safely returned home, snugly tucked up in bed, belies the statistics which indicate that a vast number of accidents occur in the home, and a sizeable proportion of deaths occur in bed at home. Naturally, these statistics must be viewed in terms of the ages of the persons concerned.

A house contains many sources of energy and of chemicals in concentrated forms – hot water supplies, steam from an electric iron or cooking, electricity and gas supplies, bacteria in drains, and jagged edges of cans and broken glass, in addition to very many threatening chemicals and poisons used in the kitchen and medicine cabinet, the garage and in garden horticulture. Falls in the home are the second highest statistic in terms of accidental death (second only to traffic accidents). Deaths from burning are at their highest in the home. Cross-infection associated with poor hygiene and poor housing is at its highest as reflected in premature deaths in the home.

These statistics are, of course, markedly influenced by the fact that one spends many hours per day in the home, that the extremities of life involving childhood and old age are often home-based, and that the home contains – on a daily and blasé familiar basis – regular risks of unsecured pans on stoves, unstoppered poisons, slippery surfaces, tortuous stairways and tripping hazards. For the elderly, a major risk of injury is the simple act of dressing and undressing and, especially, drowsy manipulation of socks and underwear.

In a typical year, an average male has a 4% chance of a home injury necessitating professional treatment, and a female has double this rate (data influenced by the fact that females spend a lengthier time in the home and kitchen and that the very elderly, over 75 years, tend to be a female-predominant group). This all suggests that we have somewhere between one in twelve and one in twenty-five chance per year (females and males respectively) of suffering an injury at home which requires treatment. There are far more persons dying from injuries received at home than on our busy roadways. The commonest causes of death are falls, which exceed poisoning and suffocation (usually food-choking) incidents. These figures, as expected, reflect that the very old and the very young contribute predominantly to these totals. The approximate annual risk of death at home is one in 8000–9000 from poisoning and other injuries; roughly 25–35% of deaths from injury occur in homes.

These somewhat depressing statistics reflect the generalization that living, be it at home or elsewhere, involves taking risks. It also reflects that the risks are higher when the person concerned has an inability to cope with the tasks demanded. This inability may arise because the industry employing the person has not insisted on adequate training and policing of the safety features, or it may be because the person is a very young child or perhaps a very old adult, all capability being worsened by the effects of alcohol or ill-health. Thus, incompetence may be equated directly to risk.

Society has a responsibility to protect these vulnerable persons from such risks because of their incompetence, lack of knowledge, or because they are cutting corners in their activities. Injury and death are too high a price to pay for not being careful. Inebriated persons, the elderly, the infirm, the sick and toddlers must all be protected from unstrengthened window glass, from poisons and burns, sharp objects, and from infection, *etc.*, all of which are encountered, to a greater or lesser extent, in our homes.

It is interesting to contemplate the role of energy in terms of risks at home. Modern homes need to be extensively energized in the form of heating, cooking facilities, washing, cleaning and entertainment features. By and large, electricity has replaced the coal fire as a source of heating, cooking and hot water. This has been reflected in a reduction in the number of burns accidents but, of late, there has been a movement back to wood-fired stoves which increases this risk.

An interesting pointer for the future is that electricity has always been regarded as a potential danger or risk and, therefore, its installation, distribution, usage and tamper-proofing against interfering small fingers has been carefully controlled, regulated and implemented by inspections, *etc.* Double insulation, child-proof plugs, fast-acting fuses and a power supply at lower voltages in North America (maximum 120 V) compared with the UK (maximum 240 V) have all reduced the numbers of tragedies. However, there is still a significant risk with any mains supply that exceeds 50 V AC.

Although town gas has been used in homes for heating and lighting far longer than electricity it is relatively recently that regulations have considerably stiffened up the protection against abuses such as suicides and incomplete combustion problems.

Thus, in a well-provided-for middle-class home that is energized from modern sources, there is a reasonable safety balance and degree of comfort for one's relaxation. Problems arise at either end of this safety plateau in that the over zealous protection of the heat in the home in the form of insulation and draught-proofing *etc.*, can give rise to a build-up of dust, toxic gases and especially radon. On the other hand, a shortage of energy can give rise to using unconventional sources of fuel and to hypothermia.

Not surprisingly, having dissuaded people from drinking and driving there tends to be a greater consumption of alcohol in the home and up to 60% of adults killed in fires in the home have blood alcohol levels which greatly exceed the legal permitted level for drinking and driving, *i.e.* their judgment and reflexes have been impaired.

Burn injuries are linked with low technology sources of warmth and with furnishings having the cheaper type of upholstering. Modern regulations now insist that all foam-filled furniture is of the combustion-modified form which tends to char or melt away from the flame, preventing the production of highly toxic gases.

A tragic score of almost 400 people die each year from hypothermia and excessive cold in the UK. Many hundreds of others succumb to illnesses which are encouraged by the coldness of their home. It is a sad reflection on Britain's energy policies that, although the overall death rate in all westernized countries is higher during the winter months than during the summer, this seasonal differential is more pronounced for the UK than for the other countries concerned.

The turnover of air within the home has been studied at length, and especially for those rooms in which the inhabitant is located for many hours, such as the bedroom and the living room. A roaring open hearth fire replaces the air in a room many times an hour. With an open chimney and no fire, there is a replacement of the air every hour or so. With modern double glazing and improved insulation of windows and doors it may take up to ten hours before the air is exchanged.

This not only influences the build-up of the range of dust, gases and odours, normally found in a house leading to enlarged numbers of reported respiratory disorders, but even cancer from the build-up of the radioactive gas, radon.

It is widely acknowledged that toxic agents are relatively harmless until their concentrations reach a threshold level. This concept refers to acute toxicity. A chronic build-up of an agent over very many years of several-hours-a-day exposure may well equal the above threshold response. Information is sparse but it is generally believed that, even if we lived in a Utopia in which there was no traffic pollution, chemical pollution or Chernobyl-type pollution in the outside world, ill effects would still arise from air pollution within our homes.

What are the sources of these threats? There are, of course, the products of heating and cooking in the form of gases such as carbon monoxide and carbon dioxide, sulfur dioxide, organic agents such as hydrocarbons and formaldehyde, the oxides of nitrogen and small particles if it happens to be a coal- or wood-fired stove.

In principle, tobacco smoke donates well over 2000 chemicals to our air and these are inhaled by smokers and non-smokers alike. Some of these chemicals are known to be cancer-forming agents or to lead to respiratory challenges.

The very materials from which the home has been constructed are now based upon formaldehyde resins used as glues for woodworking, and there is a range of dressings for the woods used in the home in order to discourage bacteria or rot. Some of these have been linked to blood disorders such as aplastic anaemia.

The threat from radon in our homes is a real one which has been tackled by the UK Government for the last decade. Radon is a gas which is given off from the decay chain commencing with uranium-238 in uranium-bearing rocks such as granite. The eventual chain terminates with the inert isotope, lead-206, but *en route* radon is liberated and, along with other gases in our air such as life-sustaining oxygen, is inhaled into our lungs on a regular basis as part of breathing. Unfortunately, this gas decays to solid radioactive products such as polonium and bismuth, and so this dust which contains alpha and beta particle emitters is embedded deep within our lung tissue where the radioactivity may initiate cancer.

High radon levels occur in the granite-bearing parts of the UK, such as Devon, Cornwall and North East Scotland, and are regularly monitored by the National Radiological Protection Board (NRPB). More recent concerns include Bristol and the Gower Peninsular in Wales which are based on limestone rather than granite, but the former rock is thought to release radon more readily. Furthermore, buildings constructed predominantly of granite and well-insulated are also a source of risk. The Department of the Environment and the National Radiological Protection Board have collectively produced information leaflets which aim to (i) prevent the radon seeping in through the floors of buildings, (ii) extract radon from under the floors of buildings, and (iii) encourage a greater turnover of the air within a home. Nevertheless, up to 6% of the annual incidence of lung cancer in the UK (2000–3000 cases per year) are believed to be initiated by the radioactive decay products of radon, this being a greater risk when the patient is a smoker.

In summary, homes are places of immeasurable benefits; from family and friends, food and drink, heating and washing, sleeping and relaxation, but there are measurable risks which need to be offset against this idealized image and which need to be reduced for certain vulnerable groups of those living in certain locations. Just occasionally, a highly laudable act, such as insulating a home, can have a side effect which needs to be carefully considered and handled in order to minimize the risk.

It is paradoxical that, because homes are such havens of comfort, the lay public are disposed to ignore a potential hazard such as radon and so governments undertake expenditure (a) to persuade the public that radiation from radon daughters in homes is potentially dangerous, but (b) that radiation from nuclear power plants is far less dangerous than envisaged!

6   Recreation

Next to sleeping and working hours, recreation is the third largest occupier of our lifetime. Psychologists have found that humans are prepared to accept far higher levels of risk during their recreation than they would be prepared to when they are at work or by living close to an industry that pollutes or poses a related risk. Recreation, of course, recharges the batteries, exercises both the body and mind, and leads to interpersonal bonding, all of which are essential to a well-balanced life. However, be it the swing of a golf club or the decibels of the exhaust of a high speed sports car, there are risks involved. One of the challenges is to decide the exact population at risk. Is it the golfer or the person hit with the golf ball? Is it the swimmer or just those swimmers that take part in diving? Table 6.1 reflects the risks generally encountered in a wide range of sports and recreational activities. What it is not able to reflect is the skill of the participant. Often, trainees or novices in a sport are far more at risk than those who practise it every day of the year, sometimes at a professional earning level. The costs of safety equipment and prevailing legislation all contribute to the overall picture.

Table 6.1   *Numbers of fatal accidents occurring in sports for males and females in a given year* (1987) *in the UK*

Type of activity	Male	Female
Air sports	20	4
Athletics	2	0
Ball sports	5	0
Equestrian riding	3	9
Motor sports	10	0
Hill sports	8	1
Cycling	2	0
Gun sports	3	0
Water sports (excluding drowning)	9	0
Playgrounds	3	3
Spectating at events	2	2

Source: OPCS Monitor DH4 88/6, Dec. 13, 1988.

Drowning is the third most common cause of death amongst all ages, accounting for approximately 1000 persons per year in the UK. There are, of course, very many people involved in water-based recreation and, as with so many other causes of death, alcohol is often implicated.

Even spectator sports are not free of their death rate. Approximately five players per year die on the playing pitch, whereas over the last decade or so, approximately 30 spectators per year have been killed in major crowd disasters. Rugby and soccer are about on a par in terms of their expected player injury and death rate. By and large, such players can anticipate approximately one injury for every two years of playing. Skiing injuries are of approximately the same incidence. Long distance running risks include heat stroke but, sadly, the risk for runners from being hit by a car is considerably higher. Three thousand accidents a year involve horse riders on our roads. The death rate from mountaineering is of the order of one in every 1500 climbers.

So why not sit at home and watch the sport on television? Although the radiation received by the viewer from the television screen has now been estimated to be at an exceedingly low level, it must be remembered that the more spectacular shots seen on the television often involve camera and film crews being located in dangerous positions and at precarious heights even though employers, by law, have to provide a safe place to work.

7 Healthcare

In this section, we do not differentiate between healthcare treatment which is provided by the medical authorities (general practitioners, hospitals, consultants, *etc.*), together with that provided by occupational health and hygiene staff in one's workplace, with those self-administered prophylactic agents and remedies acquired from pharmacies and health food shops in addition to those contributors of peace of body and mind under the general heading of healthcare. All of these actions involve risks to a greater or lesser degree and many of them produce significant benefits. The health provision organizations in the UK have done a great deal to reduce the amount of risk and to optimize the degree of benefit from, for example, an administered therapeutical, or from invasive surgery, *etc.* Nevertheless, the risk still exists and in many instances can be quantified in order to communicate that risk to the patient.

As mentioned previously, risk, which is the probability of an event occurring, may be reflected in data concerning the causes of death, the length of the lifetime before the death, or in the quality of that life. To put it crudely, we all wish to avoid premature death and to ensure that we have a pleasant and high quality life until we eventually succumb at a ripe old age!

The commonest causes of death for the older person are those of cardiovascular disease and cancer, but there is much that we can achieve in both our lifestyle and dietetics in order to reduce the risks of dying at an earlier age from these two sets of conditions. National dietary goals are directed towards reducing the incidence of obesity and of high blood pressure. The discouragement and cessation of smoking is one of the main targets in terms of carcinogenesis. However, as in so many aspects of healthcare and medicine, these two subjects are inter-related since a person who smokes ten cigarettes a day doubles the relative risk of suffering from coronary disease.

The number of patients having cancer will increase by more than 60% over the next two decades which implies more than 1000 cases per day being diagnosed in the UK, even excluding easily treatable non-melanoma skin cancer (30 000 cases per year). This means that one in two of the population will develop cancer during their lifetime (the incidence is currently one in three).

Long gone are the days when the diagnosis of cancer was tantamount to signing the death certificate. It need no longer be a lingering and painful condition terminating in death. Modern screening and treatment of early cancers have been eminently successful and epidemiology has reduced the risk of carcinogenesis arising from pollution, chemicals and other extraneous influences. Even if secondary cancer cells occur and eventually the patient does die it is now possible to use palliatives to reduce the pain, and special care hospices to ease the passing.

Lung cancer is still on the increase. Cancer of the uterus and the stomach are decreasing, the former because of early diagnosis and treatment coupled to cervical smear test screening.

Cancers of the breast, the bowel and the prostate have remained more or less constant over several decades. Although more rare than in adults, childhood cancers still persist and, in some instances, up to 90% of those diagnosed have good prospects of survival.

Part of the reduction in cancer statistics arises from recent research into the cause of this group of diseases. Carcinogens occurring as chemicals from pollution, as substances present in our diet, and from exposure to sources of radiation such as the sun or radon have all been intensively investigated and measures taken to see that the public avoid such challenges. For example, the implementation of the 'slip, slap, slop' approach to sun protection in Australia (referring to slip into a T-shirt, slap on a hat, and slop on some sun protecting cream) has benefited all of those exposed to sunlight.

One of the challenges in cancer prevention is proving causal links between an event and the incidence of a cancer because it can often take 15–20 years between exposure and cancer development. There is a related problem in respect of the clustering of groups of similar cancers. Ongoing research into the alleged, but statistically unproved, clusters of childhood leukaemias in the regions of Sellafield and Dounreay is now considering the need for at least two coincident causes to act together; for example, simultaneous exposure to a chemical and to a virus. The one certain statement in this area is that although it is exceedingly difficult to prove a causal link between exposure and carcinogenesis, it is even more difficult to disprove a claim that a suspected carcinogen is leading to an increased incidence of a cancer. To prove a negative has always been difficult in science and is even more challenging when the lead time is a decade or two before the condition surfaces. One such causal link that has been convincingly demonstrated is between tobacco smoking and cancer. This has been used to motivate organizations opposing tobacco smoking in public places (and its extremely large contribution to the government coffers through tobacco tax!).

Alleged links between radiation and cancer have been intensively investigated since the use of X-rays in the 1920s, and since atom bomb survivors of 1945, because of the large number of persons working in the modern nuclear industry and its nearby neighbours, and because of general opposition to involvement with anything that bears the word 'nuclear'. Another four-letter word (in addition to 'safe' and 'risk') which enters into the dialogue is that of 'fear'. Of all the various threats to our well-being, the fact that radiation is invisible (whereas a glowing electric fire bar or a gas flame can be seen), the fact that nuclear warheads were used to kill, genetically alter and maim hundreds of thousands of people during World War II in Japan, and that radiation can kill cells and is used as such in treating overactive or cancer cells, these have all led to the public's scepticism and fear of radiation. On the other hand, 'hospital' radiation such as X-rays or gamma-rays for diagnosis and therapy seems to be far more acceptable, partly because it is being used to treat a condition that the patient wishes to have cured, and partly because the term 'nuclear' no longer prefaces 'radiotherapy'. The public are woefully ignorant of the fact that different types of radiation can have different penetrating properties and different influences upon cells and organs, and that the length and energetics of exposure are all pivotal parameters in determining the benefits and the risks of radiation.

Similarly, the public seem to be unaware that we are subjected to substantial background radiation (approximately 2.5 milliSieverts, mSv, per annum) and that the medical uses of radiation in the hands of a skilled radiologist eliminates the need for a great deal of investigative surgery and minimizes its associated risks to the patient. The International Commission on Radiological Protection (ICRP) has estimated that, for every 10 mSv per year received by a human, the *extra* risk of dying of cancer is one in 10 000. If one examines these statistics from another viewpoint it suggests that a person aged 20–40 thus exposed will need an average of most of his or her lifetime for the development of a cancer, whereas a person aged 80 or 90 has very little time left in which to develop the cancer. Once again, the latent period between exposure and the first cancer cell appearing complicates the issue.

To place the figures on a UK basis, if 1 000 000 of the population received a 10 mSv annual dose, then a statistical figure of 100 additional deaths could be estimated. Naturally, different organs and different persons of different age groups are more susceptible than others. The current chances of dying of cancer in Britain are between 20 and 25%, *i.e.* approximately one in five. Thus, a 10 mSv dose over your whole body will add an extra chance of one in 10 000 which raises the death figure from an average of 22.50% to that of 22.51%, assuming that you are in your middle life. The NRPB some ten years ago indicated that the average exposure in the UK is now 2.5 mSv per year from background radiation. The increased risk of cancer occurring when exposed to an extra 1 mSv is that of one in 100 000.

Another potential danger from irradiation is that of congenital abnormalities arising from irradiating the foetus within the expectant mother. Other abnormalities are those which are genetic changes brought about and inherited throughout a range of generations. The unfortunate victims of the Hiroshima and Nagasaki bombs and their grandchildren have been able to contribute substantially to these investigations. Fortunately, it appears that the radiation-induced mutations have a very low incidence and are rapidly bred out of a family's genes.

The overall figures suggest that background radiation may well be linked to 1% of all cancer deaths. Man-made radiation sources, as used in hospitals, *etc.*, might account for 0.1% of cancer deaths, and radiation from nuclear activities, at less than 0.01% of background, may therefore contribute less than 0.0001% to risk of cancer deaths. Nevertheless, it is that last low fraction that attracts the attention of the media and protest groups.

8   Diet

The role of diet in health is the subject of intense research and marketing activities. Roughly one third of one's diet is good for health. One third has something good about it. The final third could well be contributing to undesired characteristics, such as obesity, cardiovascular disease, *etc*.

For the carnivores of our forefathers, chasing, catching and devouring prey along with a liberal dose of contamination from topsoil, *etc*., this might be regarded as having eaten a fairly healthy diet except where the meat was contaminated with some undesirable bacteria. It was high in fibre, low in fat and did not have added sugar, salt, milk or sweeteners. It would also have been rich in polyunsaturated fatty acids. Traditional cooking techniques whereby meat and vegetables were taken straight from the soil into the cooking pot and stewed for an hour or two before being eaten had many advantages over the modern, fast-food society, which depends heavily upon convenience meals which may be prepared during the commercial breaks on television.

Furthermore, no longer needing to hunt for one's prey, but rather enjoying one-stop shopping at the local hypermarket, leads to lack of exercise and obesity in some. Many adults now weigh more than 20% above their desirable weight range, and 50% of men and 49% of women in the UK are now obese. Although we are eating 800 calories a day less than we did in the 1950s the proportion of fat in our diet has increased by 50%. Further, 60% of adults do not engage in physical activity. This situation makes the person prone to illnesses and conditions such as gall bladder disorders, coronary heart disease, high blood pressure, diabetes and cancer. Sir Richard Doll has estimated that up to a third of all cancer deaths in some countries may be attributed to dietary factors.

Modern food technology also influences our health. The ability to be able to process food, preserve it for long periods of time, to achieve flow during the mixing process, *etc*., and to stimulate the taste buds means that modern foods have different bioavailability (uptake from diet) and fibre characteristics than traditional foods. A recently launched fat substitute, or rather a modified fat, retains the pleasurable taste and texture of natural fat yet passes along the intestine unabsorbed, contributing negligible calories. It ostensibly attracts vitamins, nutrients and cancer-protecting antioxidants and is linked to physical side effects. We are a long way from following Grannie's advice to 'eat your vegetables, go outside and enjoy your play'! On the other hand it must be pointed out that it would not be possible to feed the World using traditional methods of hunting and gathering.

Finally, there is the comparatively recent introduction of psychology into our eating habits in Western Culture where famine is not a major problem. Menus now reflect the social and symbolic status of food.

9   Medical

A typically modern challenge concerns food irradiation, as part of longer
shelf-life technology, to inactivate micro-organisms using the shortwave
ionizing radiation of gamma-rays, X-rays, *etc*. Although modern dietary
habits and food processing may change the amount of essential vitamins
in a food, the public's replacement of these vitamins from health food
shops and pharmacy outlets may not completely restore the balance. In
terms of nutrition with trace elements and with vitamins, it is well to
remember that a slow leak can eventually sink a big ship. All of us are
sometimes ill or take medicines as prophylactics to prevent serious
disease. This means that we are all committed to medicine and,
occasionally, to surgery. Although patients tacitly assume that all
medical attention has some risk associated with it, they tend to over
simplify the information and assume that all medicines are good.
Surgeons involved in an operation, doctors and patients alike all aim to
maximize the benefits and minimize the risks. All drug designers set
about designing a risk-free therapy but, in fact, there is no such risk-free
drug and even the safest of drugs taken in overdoses can be seriously life
threatening.

Part of the approach to reducing the risk is to listen carefully and
attentively to what the medical doctor or the surgeon advises the patient.
However, studies have recently shown that only one-fifth of the information imparted by the specialist is retained by the patient during their
period in hospital (see Section 11). This memory lapse potentially
increases the risk and does not maximize the benefit.

In the ultimate account, treatments which carry a finite risk can be
judged from the death rates, but medical statisticians gather far more
data from undesirable side effect surveys and from closely controlled
clinical studies. Thus, it is well known that there is a risk from surgery if
an anaesthetic is used, either local or general. Currently this figure is
approximately one death per 25 000 operations.

Babies, up to 2000 deliveries per day, are now born in hospitals with
just a minor proportion born at home by intent. The death rate for
mothers is now at a low level of one per 10 000 childbirths.

History provides several instances where the procedures have now
been shown to carry more risks than the benefits they are assumed to
produce. A gall bladder removal operation rate five times higher in some
countries than in other comparable countries has been linked to deaths
due to gall bladder disease being double in the higher operating country.
Tonsillectomy of some 30% of children before their middle teens did not
apparently reduce the rate of ear and throat infections but did produce a
significant number of deaths from the operation. Recent data indicate a

similar negative effect in terms of morbidity as well as mortality in that there are negative psychological effects of health screening which produce a false 'positive' result and cause depression long after the test has been repeated.

Under serious life threatening circumstances, a very high risk of death can still be the preferred way forward for the majority of patients. For example, *emergency* brain operations carry a risk of death of one in ten and, similarly, for heart and other chest operations a risk of one in thirty. At the other end of the life threatening scale, tonsils and piles operations carry a death rate risk of one in 10 000 operations.

A far more common experience is that of taking medicines. The life expectancy at birth has increased markedly this century because of the use of extremely effective medicines and vaccination techniques. Nevertheless, disasters do occur and some 35 years ago 8000 children were born deformed because of the mis-prescribing of the drug thalidomide. Toxic and teratogenic effects may be identified from selective animal experiments and from consenting-adult clinical trials. Side effects thus identified can sometimes be reduced by modifying the formulation or the means of administration, but reports from a very large number of patients are necessary in order to prove convincingly that a drug is causing side effects over and above the normal scatter of opinions and ups and downs of the everyday health of an average person.

Much of this problem arises because the side effects may not be immediate. Similarly, the reporting mechanism from patients taking the drug through their GPs may not be highly efficient. It can be shown statistically that, if the incidence of an adverse side effect is one in 500 patients using the agent and if the normal background risk of that side effect, such as headache, pain, *etc.*, is one in 100, then a minimum of 36 000 patient reports from GPs to drug producers would be necessary in order to indicate whether the new drug does indeed increase the incidence of the side effect.

When answering questions from patients concerning whether the drug is safe or not, doctors have to be aware of whether the drug will be used on a long-term basis or whether it will be short-term therapy. A drug which exhibits a risk of dying of one person in 10 000 and which is administered for a limited period to just a few hundred patients per year may be regarded as a very low risk, whereas if 2 000 000 patients took that drug regularly on a daily basis for 20 years of their life, there would be a large number of deaths attributed to the agent concerned.

All of these statistics are now being distorted by the fact that the reporting of adverse side effects of drugs is a voluntary matter and, also, that there are now agencies who make vast profits out of suing drug companies and doctors for the side effects of a drug. Approximately

16 000 or 17 000 reports are received each year concerning side effects, and many of these concern skin rashes arising from therapy. Unfortunately, these are not the figures which are reflected in the headline news reported by the media. Journalism, when irresponsible, causes much suffering, and unwarranted fear of rare side effects can thus throw the baby out with the bath water.

Such unwarranted fears have arisen concerning vaccination of children. The balance between a disease which occurs if the child is not vaccinated and the side effects of vaccination is a classical dilemma facing parents and family doctors. No matter how risky a vaccine, it is axiomatic that the illness is always more dangerous. Twenty years ago the incidence of death from whooping cough (pertussis) vaccination was one in one million. Brain damage occurred at ten per million and a set compensation payment was arranged at £30 000 if the child became 80% or more disabled.

Measles vaccination and side effects have displayed similar trends from some 70 per million vaccinations. However, just recently, measles-vaccinated persons using the MMR (measles, mumps and rubella) jabs, are reported to show three times the expected incidence of bowel disorders, and DPT (diphtheria, pertussis and tetanus) jabs are reported to produce three times the incidence of seizure.

The use of X-rays for both diagnosis and treatment is an interesting illustration of weighing up the risks and benefits of this irradiation-based procedure. Early experience with the technique, which was invented in 1895, indicated that doses which were too high could clearly lead to cancers in organs such as the breast. Widespread screening using X-rays at one time caused more deaths than it prevented by early diagnosis. However, modern, low-dose irradiation appears to have an almost minimal cancer-forming risk in persons aged 50 or over.

In fact, medical and dental X-rays form the largest proportion of artificial radiation administered to the population at large. It must be remembered that every X-ray carries some risk, albeit small. A normal X-ray is of the order of 0.7 mSv, which is the same risk as smoking five or six cigarettes. For a 40 year old person, this risk is associated with one in 250 000 deaths, which is the approximate risk for a person dying of natural causes on any one day of the year. There are X-ray procedures which involve greater doses and risks, such as barium meals and barium enemas. The overall impression seems to be that medical procedures involving irradiation are becoming less risky but that there is a need for more informed information to be given to the patients concerning the relative risks concerned.

A related comment is that the absence of information does not imply zero risk. Although the risks from the various medical procedures previously mentioned have been stated, the fact that many natural and herbal remedies have not had a scientific risk assessment reported ought not be taken to imply that they are necessarily safe.

10 Energy

The Domesday Book of 1086 AD mentions the dangers of working near to the mechanism of water mills. In addition to hydro-power, which has been known for thousands of years, so too is fire-power, in the form of burning wood and peat, a source of risk from toxic gases in the smoke. Modern versions in terms of wave power and hydroelectricity, and fossil fuel and oil combustion also have associated risks, as do other forms of power such as thermal geyser power and nuclear energy. Nevertheless, modern legislation in terms of the Health and Safety at Work Act and Duty of Care for those persons likely to be affected by a power station have substantially reduced the risks to low levels of contamination.

Western inhabitants require the order of three kilowatt hours of electricity from the moment of birth to the moment of death in order to produce their heating, lighting, transport, clothing and accommodation, and all the other accoutrements of modern western life. This demand is a factor of ten higher than early humans required from their campfires. The developing world is rapidly catching up in terms of its energy desires and, when one considers that the World population is approaching six billion, one realizes that energy demands for the future will exceed current day capabilities worldwide in spite of serious conservation and recycling measures.

Another feature of our energy requirements is that we prefer our power to be instantly available, which means that supply lines or tanks of energy are juxtapositioned to our cooking appliances, motor cars, *etc.* Such energy stores in the form of dams or high voltage power lines obviously have an inherent risk in the event of failure. Further, they run the temptation that similar technologies can sometimes be used to release warmongering devices against enemy lands. Thus, water-borne dysentery was used as a weapon against the Romans (hence their preference for sanitized beer and wine to drink!), smoke screens were used in both World Wars, nuclear warheads were used on Japan, and fire has long been used to burn out enemy strongholds. These points are important when one comes to the perception of the risk involved. It must be stressed that, for public acceptance and in order to maximize the production and usage of energy for peaceful, household, domestic and industrial outlets, conditions must be entirely different to those used for concentrating the energy in the form of a warhead. The two atomic technologies have many differences and a nuclear power station cannot explode like a nuclear bomb.

Opponents of a particular form of energy production or of its location in a particular area will use selective arguments to make their point and often they compare risk values which really ought not to be compared. It is important to compare the whole cycle of energy production from a particular source, and this includes the mining of the raw materials, through to their processing, usage, recycling, if necessary, and the disposal of any wastes into the geosphere or atmosphere. Similarly, the risks to operators of such a power station must also be taken into the calculation.

Most countries, and the UK is no exception, have a portfolio of energy sources which includes fossil fuels such as coal and oil, natural gas, hydroelectric power and nuclear energy. Dependent upon world markets and the non-accessibility of certain categories under certain conditions, this portfolio is modified and prepared for the future minimization of the cost of energy. A nation's industrial production depends upon a goodly supply of cheap energy. An increase in the price of electricity or gas does not just bear the risk of increased prices to the household consumer, but also bears the serious risk of closing down large energy-intensive industries with related job losses having safety and health implications. Even the cost of constructing the power stations will have a risk cost in terms of life. Modern quantity surveyors and statisticians can roughly predict the number of expected deaths per £10m spent in terms of a steel construction, or pouring concrete or working out at sea on a North Sea oil rig.

Tremendous progress has been made in terms of reducing the risk from energy production. Some 30 years ago, half of the fossil fuel used in the UK was burned in the open grates of private houses and gave rise to choking smogs accounting for hundreds of deaths per year (approximately 4000 deaths in December 1952 from a typical London smog – this culminated in the Clean Air Act of 1956). The pooling of such combustion in modern fossil fuel-fed power stations and the supply of energy in the form of gas and electricity has markedly reduced this annual event of the London peasouper smog; even so, in 1995, faulty gas appliances and blocked chimneys were responsible for 63 carbon monoxide deaths.

Even the production of the raw material involved sacrifices. An average of 500 miners died per year in the middle of this century but this figure is now down to an average of teens. Thus one can quote an approximate risk of one death for every 2 000 000 tonnes of coal mined. Open-cast coal mining and a change of philosophy in terms of underground safety has led to this improvement. Similarly, one must calculate the health cost to uranium miners from the radioactive yellowcake which goes into their lungs. These deaths may be acute, as from a roof fall in a mine or long-term chronic deaths as with the pneumoconiosis associated with miners' lungs. Not only do they have this terrible condition to bear but also there is the increased damage done by the radon from within their mines and the X-ray procedures used in its diagnosis.

Off-shore oil and gas rigs also have their fatalities, and a figure of 50–100 deaths per year was not unexpected during the times that rigs were being constructed and commissioned. Nowadays the fatalities from uranium production per kilowatt hour is about one-tenth of the mining deaths of coal production. One spin-off of studying the death rate from lung cancer of uranium miners has been a closer definition of the risk associated with radon in houses.

An interesting calculation concerns the waste arising from uranium mines. These 'tailings' produce radon and thorium gases from the decay of uranium-238, and these pose a similar threat to radon in homes if entrapped and enters into the lungs. However, if the tailings are covered with an impervious layer of earth or rocks, emissions may be decreased; but, unfortunately, the half-life of radium-226 is 1600 years and that of thorium-230 is 80 000 years. A rough estimate of the fatalities arising from this surface disposal of uranium mine tailings is of the order of one death from cancer for every ten years of operation of a nuclear power station running on uranium fuel rods. There is also the question of whether the radon and thorium contained and trapped within the tailings will eventually burst through within the next 80 000 years or so.

Both coal and oil require transport and this has an associated small risk. A larger risk is envisaged from the movement of oil, and more recently of liquid petroleum gas tankers on our roadways, and it has been estimated that about a dozen deaths per year arise from collisions with these juggernauts.

Coal-fired power stations emit toxic gases and particles including tiny droplets of sulfuric acid. Oxides of carbon, sulfur and nitrogen are all emitted in vast quantities. This means that people do not wish to live in the vicinity and this creates additional risks from transport in driving to work. The gases emitted from a coal-fired power station contain carbon-14 and also radon, and this often means that the radioactivity emitted from the stacks of coal- and oil-fired power stations is considerably higher than from comparable nuclear power stations. Incompletely combusted carbon compounds emitted include benzpyrenes, which are carcinogens. Finally, the fly ash from coal-fired power stations contains radioactivity and is also present in large quantities thus requiring disposal. Its acidic nature means that it is unsuitable for many construction purposes.

Our nuclear power stations, by contrast, have many layers of safety protection facilities built into their design and the more modern ones have routes for decommissioning at the end of their useful lifetime designed *ab initio*. This latter point will significantly reduce the wastes arising and also the operative exposure to radiation involved in the decommissioning process. The main threat from nuclear power stations is one of perceived risk in the form of their linkage to nuclear bombs (although this is not feasible in practice) and, also, the assumed link between radiation and cancer. In fact, the larger risks of radiation exposure are in fuel preparation and, more significantly, the fuel reprocessing aspects of used fuel rods for a nuclear power station. In the UK reprocessing is done at British Nuclear Fuels sites such as Sellafield, *etc*. Undoubtedly, the most dangerous material in the nuclear cycle is that of spent fuel, and a series of fail-safe procedures is used to ensure operator protection from fuel rods when defuelling the power station, reprocessing the spent fuel, and disposing of the highly active wastes (HLW) therefrom.

Workers in nuclear power stations who may be subject to radiation from beta and gamma sources are all carefully monitored with personal and room monitors, and regular medical check-ups. This punctilious and careful approach is evident on a visit to Sellafield where only a very small quantity of short-lived radioactive waste arisings goes down the pipeline to the Irish Sea. These liquid discharges have been reduced by many orders of magnitude during this last decade by the expenditure of up to £1bn on effluent clean-up equipment in the form of ion-exchange plants, *etc*. Nevertheless, some isotopes become part of the food cycle and humans living in the area eating laver bread or fish contaminated from this source form a critical group of persons who are carefully investigated in order to determine the worst possible case intake and daily distribution of isotopes arising from Sellafield. Even these extreme, critical group members do not reach 50% of the permitted levels established by the International Commission on Radiological Protection.

In the 1980s, the incidence of groups of childhood leukaemias in the vicinities of the Sellafield and Dounreay nuclear plants led to extensive searches involving groups of experts from a range of disciplines for causal correlations. In spite of thorough searches involving many different routes, there still remained an unexplained safety margin of at least 250–300 times between the maximum amount of irradiation that could conceivably have escaped into the parents of these children and the number of cancers that might have formed. These groups of leukaemias were never statistically sufficient to be called a 'cluster' in mathematical terms. Nevertheless, the Committee on the Medical Aspects of Radiation in the Environment (COMARE) recalculated the risks of leukaemia from Sellafield and all other major nuclear establishments, such as

Aldermaston and Harwell in the UK, and failed to find the origins of these childhood cancers. Searches for other 'cluster-like' groups in the UK revealed more than 500 sites, the majority of which had no nuclear facility in the vicinity whatsoever. Nuclear installations did not have state-of-the-art monitoring equipment at the time of the releases in the 1950s and so data from that era were inadequate to track down any causal link from such low levels of radiation ostensibly emitted at that time. Modern monitoring and health physics facilities are, of course, providing extremely reliable data for today's cluster-seekers.

There was also the suggestion that a combination of two or more factors may have triggered off leukaemias in these areas in the 1950s. The 'Herd immunity' theory involved isolated groups not having built up their immunity to disease brought in from suburbia when new technologies arrived in the area; alternatively the origins could be a combination of chemicals from nearby chemical plants and radiation, or viruses coupled with radiation or chemicals. These have all been considered but, as yet, not proven as causal links. The jury is still out!

Much of the foregoing discussion of risks of various operations involves normal, everyday procedures in which major incidents have not occurred. However, many readers will have heard of the explosion of the Three Mile Island nuclear reactor in the USA in 1979 and all of us must surely have heard of the 1986 disaster at the Chernobyl nuclear reactor in the former Soviet Union. Both of these released large amounts of energy and proved that human error must be anticipated in the design of future power stations.

It was a deliberate operator decision at Chernobyl to switch off safety features such that the reactor was driven into an overheating situation from which it could not recover. At Three Mile Island, large safety labels tied on to switches on the control panel covered over red warning lights which ought to have alerted the operator to a steam build-up.

The coal industry has had many disasters, both underground and at the surface. In the 1930s the Gresford colliery disaster in North Wales accounted for 250 dead and the Aberfan, South Wales school disaster, when a coal tip moved under pressure of rainwater, killed 144 children in 1966.

Hydroelectricity and its requisite dams have killed many thousands of people. The largest ever death toll from power generation occurred in India in 1979 when the Gujarat Dam burst and killed at least 15 000 people. Over the last half century, more than 100 power-generating dams have collapsed in the United States alone. The Peoples' Republic of China is currently constructing a dam three times larger than any dam existing anywhere on this planet. The death toll in the event of this dam structure failing could be immense.

Oil can lead to environmental disasters when tankers run aground and break up releasing their cargoes or if they ignite; for example, a liquid petroleum gas explosion in Mexico City killed more than 500 people.

The radiation emitted from Chernobyl was approximately 100 times worse than the fire at Sellafield (formerly Windscale) in 1957 and approximately 1000 times worse than the Three Mile Island occurrence in 1979. This somewhat reflects the different emphasis upon safety and procedures used in different countries. Certainly, the gas-cooled reactors such as Magnox and AGR used in the UK would not have overheated were there to be a failure in the power supply to the cooling circuit (this was the experiment involved in the shut down at Chernobyl). Multiple automatic shut-down devices would have extinguished the reactor long before the temperature rose. Another feature is that of emission and steam containment, which would have ensured that there was no release of radiation. The question arises as to whether it was a system accident or whether it was human error. In fact, the current view is that it was the fault of the reactor operators in combination with the reactor designers and, indeed, with the administrators of the Soviet nuclear power plants, all having committed serious errors. The operators had been insufficiently trained in reactor physics to be totally aware of their actions. The plant should have been designed to be fail-safe. The government and nuclear administrators should have built unambiguous responsibilities into the legal regulations for the use of nuclear energy and for the assurance of nuclear safety.

Of the sources of energy which have been investigated and used, solar energy, wave power, wind power and energy from renewable organic sources of biomass are well known. Risk calculations have not been as extensive as for the nuclear energy field but it suffices to state that the raw materials required for the technology and the concrete buildings are of a similar order of magnitude to those used for coal, oil, gas, hydro- and nuclear power. Interestingly, the energy required to produce the photovoltaic cells used in solar power stations and for purifying the materials therein is a substantial outlay. Once in use, they are susceptible to interference from dust and clouds, *etc*. The installation of windmills at the top of towers or of solar panels on the rooves of houses all involves risk. Most smaller, localized sources of energy require a back-up system, which is often a petrol-driven generator, or storage of electricity in lead–acid batteries, and both of these are statistically high risk technologies, especially when only used by an untrained person on rare occasions. Generators located out in the deep water in order to use wave power generate risks for servicing personnel and for navigation as well as damaging the ecology. Windmills, once established, require less servicing but take their death toll on birds attracted by the wind currents and by moths, *etc.*, to be found thereabouts. If one wishes to produce alcohol from fermentation of sugar cane, the agricultural practices used to plant and harvest the sugar cane are some of the most dangerous worker practices in any industry, since agriculture is definitely a very high risk occupation.

The UK Atomic Energy Authority commissioned a study into the relative risks of the principal sources of energy generation. In spite of a wide margin of error in the calculations concerned, this study by Ferguson reported that all three of the major fuel cycles, nuclear, oil and coal, are associated with low risk estimates, bearing in mind the large amount of electricity that they produce. Because of uncertainty overlap, it was not possible to say that one was preferable to the other. This is not diplomacy, but rather is typical of risk comparisons when highly complicated and not terribly comparable risk scenarios are considered over a long time span between commissioning and decommissioning.

The overall conclusion from these studies is that switching from one type of generator to another will not be expected to take us forward a quantum leap in terms of reducing current risks which are already exceedingly low. Rather, decisions in the future will be based upon the relative costs of different types of energy production and, more importantly, by the perception of the public about these more or less equal risks.

What about risks to future generations? The Radioactive Waste Management Committee of the Nuclear Energy Agency (OECD) has addressed the question of sustainable development put forward by the World Commission on Environment and Development (the Brundtland Commission) of 1987. This is defined as 'satisfying the needs of the present without compromising the ability of future generations to meet their own needs'. It hinges upon the question of what is acceptable as morally correct human conduct. Major questions arise such as intergenerational equity and intra-generational equity. This question is also addressed by Sir John Knill in his lecture, entitled *Radioactive Waste Management in a Sustainable World*, where he also writes about the 'precautionary principle' to environmental decision-making.

11 Perception

Life is a group activity and most people belong to several groups. There is a work group, a professional grouping, a family group, a group of neighbours in the location, and a political group, all of whom express their views.

In a society which claims to be democratic, there are many different types of group involved in collective decisions which affect many areas of society by imposing certain risks and benefits upon them. For example, society has to decide whether to build a by-pass, whether to fluoridate water supplies, whether to distribute school milk or to build a waste disposal site in a given area. It has been demonstrated over many decades that each of these groups will perceive the suggested development differently. In principle, they can all be given access to the same amount of data and statistics concerning risks and yet, human nature being what it is, different groups will recommend different ways forward depending upon whether they are politically involved, whether they are the technological experts involved in the design and planning of the event, or whether they are other group managers.

Let us assume that there is a proposal to build and develop, or to implement, a community beneficial scheme many hundreds of miles away and definitely 'not in our backyards'. It is axiomatic that, were we able to persuade each of the members of the groups concerned in our locality to consider all of the data and statistics carefully, the overall recommendation made could well be totally different to the same scheme being proposed for an area within our perceived backyard. The NIMBY principle (Not In My Back Yard) is a great leveller of opinions! Initial gut reaction to any proposal tends to respond, in the first instance, 'What is in it for me and for my group?' On the one hand, the public has grown distrustful of politicians because they are suspected of having ulterior motives. On the other hand, experts can sometimes be dismissive concerning passing on their knowledge and expertise to the general public, or else they simply make it difficult to understand.

The perception of a risk means entirely differing outcomes to different persons. This section describes the history of the perception of hazards and risks, of how data concerning risk enters into the decision-making process, of how influence is brought to bear by introducing certain dread factors, of benefits to members of the public from accepting a new risk in their area, of the role of statistics and, indeed, of the presentation of risks to a wide audience. Only through full appraisal of all of these aspects will it be possible to introduce risks which are acceptable and to use risk management wisely in the decision-making process.

One of the first features to put across is that each of us is subject to different risks and that the quality of life, and indeed our progress through life, only occurs because different members of our society are taking different risks for different reasons. For example, the quality of life of living in some remote, environmentally idyllic spot in the Highlands must be counted against the risk of an ambulance not reaching you in time in the event of a crisis. Similarly, the first class quality imported fruit and foods in our larder has involved shippers and stevedores taking risks delivering these cargoes on our behalf.

It is salutary to ponder that, even if we could invent a process for turning iron ore into gold, there could well be groups that opposed the establishment of such a "philosopher's stone" factory in their locality because of the perceived threats of construction, and of factory noise, or of it attracting robbers, *etc.* Thus, not surprisingly, when attempting to acquire planning permission for a waste disposal complex, one must anticipate less than enthusiastic thoughts from the local community.

Individuals perceive risks differently from each other. Groups could well have a group compromise perception of a risk which differs from that of any of the individuals in the group. Long gone are the days when decisions were fairly straightforward. For example, "I am hungry, therefore I will hunt, kill and eat and if a mishap occurs and I get injured, that can be blamed on the deity rather than myself for having hunted too close to a cliff precipice or too near to a river's edge." Thanks to modern communications and education, we are now made aware of the dangers of activities such as food gathering and we can also benefit from the experiences of neighbouring groups or tribes. Nevertheless, the decision to cross a road away from the established crossing point, or to hunt in the wild is a personal decision which is under our own voluntary control. We introduce highlights into our lives by taking decisions that are perceived to be more risky by, for example, participating in some sports.

Last century, with the coming of the industrial revolution, risks became collective and new hazards emerged which were imposed upon societies by group decisions taken by managers located elsewhere. No longer were most of the risks facing our lives ones of voluntary control. Decisions introduced new hazards into our lifestyle over which we had no input. Sometimes, we first heard about them through headlines in the media, and so it is not surprising that they were not welcomed with open arms by the individual concerned! Psychology researchers have produced many theses concerning perception of the risk, and have even developed mental strategies whereby they can manipulate a person's impressions through the well-planned order in which the information and statistics are placed before them. To be frank, the media skills of manipulation of perceived risk scenario in terms of increased media sales through 'shock-horror-scare' headlining and selling films through terror of the unexpected are far more advanced than those of technology developers laying out information for interested parties in order to win them over by a series of simple decisions. Later we will discuss how scientific spin doctors can be used to allay fears partially in terms of perception of risk. One of the interesting features of this is that if one over-stresses the safety measures used to prevent a particular incident ever occurring, the public tend to be alarmed that such safety measures are necessary. It is seen as implicit that there is a far larger threat lurking somewhere in the background. Similarly, the perfectly correct insistence that all incidents of a potential risk within a large organization should be made public (for example, any leakages of radiation and even cut fingers occurring in a nuclear establishment), and then the summation of these reports indicating that there were a dozen or so very minor non-threatening incidents over the last few years, is construed as meaning that accidents occur frequently and that the next one could well be a major incident.

Much research has gone into the perception of judged and actual threats as reflected in death rates and suchlike. Similarly, females tend to rate risks approximately 10% larger than do males (see Figure 11.1). X-Rays are regarded as being of low or negligible risk because they are 'medical' X-rays. Similarly, the use of nuclear magnetic resonance imaging of brains in hospitals has been considerably more popular with patients and relatives since the word 'nuclear' was dropped. By and large, the public have high expectations of scientists and medics. Since science has been so successful in allaying many of the killer diseases which affected our populations at the beginning of the century and in providing a plentiful supply of fresh water, clean air, electric power, *etc.*, it is assumed by the public that they can also take away all other threats and the associated perceived risks and, in the extreme, to produce a totally healthy life which lasts for infinity! Another feature is that the scientists

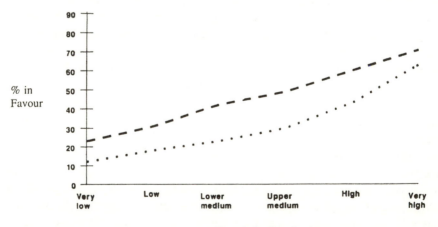

Figure 11.1   *Perception of the risks of nuclear power expressed as percentage in favour for males (– – –) and females (· · · · ·) across groups of subjects having different intelligence quotient*
(Source: RSL)

are not held in high esteem by politicians because we are not the best of communicators and do not go to meetings prepared to give a straight answer to the simple questions such as, 'Is it perfectly safe?' and 'Will it lose me and my Party votes?' Whereas politicians deal in certainties, scientists research probabilities!

The dread factor involved in the choice of terms used to describe conditions and risks has a significant influence upon perception and can often tilt the balance against a situation. For example, if fairly accurate statistics were given concerning fatality rates from a variety of diseases at the age of 70 years old but known only by a given code number, the public would fear the risk from the highest percentage of deaths. However, as soon as a code number is replaced by the word 'cancer', regardless of where cancer ranks in the order of death rate, it will immediately rise to the top of the perceived threat index. On the other hand, if a group of deaths was attributed to deaths from excessive sun-bathing, or social drinking, these would be ranked far lower in the perception stakes. This suggests that exposure of the public to certain terms and certain concepts and their familiarity with the origins of the condition dilute their image of a given risk. The public who are not expert in a particular discipline perceive according to their subjective evaluations as influenced by the press or activist groups rather than by cold data.

Research from academics in the UK has suggested that members of the public prefer to base their estimates of risks to themselves upon two pieces of information: the first is the likelihood of a mishap occurring, and the second is the likelihood of that mishap leading to their death. Exact details concerning numbers and the properties of the group exposed to this mishap do not seem to influence their personal perception of the risks and benefits of a proposed measure. In this last respect, it is more difficult for members of the public to comprehend the perception of the risk than the perception of the benefits. The logical outcome of these tenets is that we are prepared to make extremely expensive rescue missions for a lone person trapped with very little chance of survival, whereas a fraction of that outlay used in a constructive manner to provide safety measures covering a year or so could well save a large number of lives. If the risk is one of a self-imposed, voluntary nature, usually because of some pleasurable activity or sport, then the perception of this risk is one of acceptability rather than of threat. The corollary is that an imposed risk from an outside body which is hundreds or thousands of times less serious than the voluntary activity will be perceived as being a serious threat and opposed with vigour. This lack of personal control over one's own destiny in terms of risk is a pivotal determination of risk perception.

A recognized problem with the perception of the public is that persons imagine that they are less at risk than the average member of the society. Weinstein called this 'optimistic bias'. It is apparently a subconscious desire to feel in control of a situation. Hazards that are more easily controlled have more optimistic bias, for example, smoking, alcohol intake, *etc.*, whereas those under the control of others, for example, additives, herbicides, *etc.*, do not have an optimistic bias.

This then brings into court the question of "accepting the Queen's shilling", *i.e.* I would be prepared to volunteer for any risk provided that the financial benefits to me were sufficient. This all suggests that large organizations involved in exploration, energy generation, waste disposal, healthcare, *etc.*, should place more emphasis upon making known the benefits to the persons affected by their proposed operation rather than just stressing the cold statistics of the extreme unlikeliness of an incident occurring. They should perhaps also place more emphasis upon the likely damage to the person concerned if the most unlikely incident occurred. In many respects this damage will be almost zero but this point will be better received if made at the planning stages rather than after an incident has occurred.

Thus, so far, we have indicated that perception is based more upon voluntariness, benefits to the person concerned, likelihood of an incident occurring, likelihood of damage to the person concerned and first-aid fire-fighting provisions, rather than numerical data. Coupled to this list must be the image that the person perceives concerning the proposed activity, be it a surgical operation or an industrial development. So often these images are determined by the media. For example, television programmes regularly attract audiences of many millions if medicine, hospitals, crises, emergencies and drama are involved. Yet to quote the words of John Cleese, 'Mention the chemical industry on television and a million people immediately get up and make a cup of tea.' Perhaps more could be done to point out the benefits to all concerned arising from products of the chemical industry in terms of healthcare, food, pharmaceuticals, overseas trade, finance, *etc.*

A peculiar, reciprocal relationship exists concerning safety measures. The more safety measures that are introduced into a system and the more frequently they are tested and shown to work efficiently, the more the non-technically minded observer perceives the threat of the industry to be real and serious. To the technologist, however, this array of safety measures under efficient operation is viewed as a successful outcome. The different reciprocal viewpoints can be seen by optimists and pessimists claiming the bottle is half full or half empty, by quoting the death rate or survival rate, by quoting the expected life expectancy or risk of death, *etc.* It often seems that the media, when selling their products, prefer to use the approach that every silver lining has a black cloud in front of it! It is the threat of this black cloud that enhances their sales and profits!

Another serious obstacle to be tackled with a lot more effort than that currently made is the fact that comprehension research has shown that members of the public facing challenging decisions, such as whether to have a serious surgical operation, or whether to go on a climbing holiday in spite of their bad heart, *etc.*, only retain approximately 20% of the information that they are given by professionals. This retention is judged at the time when they deliver their decision to make the booking or to sign the consent form for the operation. Some medical authorities are now insisting that consultants presenting data to patients for them to contemplate, sleep on and deliver their decision by the following day must present the information in written form. In the rare event of the operation not being an outstanding success, the human mind tends to be selective with its residual knowledge and forms a biased opinion (possibly encouraged by ambulance-chasing lawyers) which could well misrepresent the surgeon and colleagues.

Paradoxically, the public has an ambivalent approach to being definite. On the one hand, they are highly sceptical of an organization which implements many safety checks and measures and fail-safe devices in order to almost guarantee the safety of a piece of equipment or technology. On the other hand, people perceiving risk wish to have definite assurances quoting odds rather than just qualitative terms. Clearly, the more one is expected to be definite, the more checks, monitoring and data collection exercises are necessary. Treatment manufacturers and the pharmaceutical industry are well aware that it is necessary to present their healthcare agents in the correctly perceived light. Pain treatment is described as 'relief of pain'. On the one hand, scientists and statisticians can be reasonably objective in terms of defining a risk and the likely outcome of an occurrence. On the other hand, lay-persons and members of the public are far more subjective in terms of assessing that risk. Because the latter group obviously have a vested interest in discussing the question, their views of the perception are equally as valid as the professional view based upon statistics. There are those who believe that feelings are sometimes more important than numbers. Thus, arguments professionally presented based upon numbers and risks ought not to be expected to change a lay-person's opinion of their perceived risk. There is an important gap to be bridged in terms of communication and knowledge transfer. The British Medical Association in the UK has stated that, 'Knowledge does not ensure that decisions will reach universal approval, but without knowledge, the extent of disagreement and disapproval will almost inevitably be greater.'

There are two important messages from this discussion of perception:

(a) Professionals involved in risk assessment are under an obligation
    to go that extra mile and take on board the lay-person's, albeit
    qualitative and most definitely subjective, views of their data.
(b) Lay-persons need to be persuaded by technologists that there
    really is a need to learn to understand data and that the more data
    that they understand, the more rational and reliable will be the
    basis of their decision making.

Researchers at the Stockholm Centre for Risk Research have investigated risk perception in terms of psychometrics and cultural theory.
Factors responsible for the public's perception of risk may be considered
as:

(a) the benefit to society;
(b) the perceived risk;
(c) the acceptability of the current level of risk (this gives a risk
    adjustment factor: how many times safer should it be, or how
    many times riskier could it be?);
(d) nine risk dimensions:
    (i)   voluntary or involuntary;
    (ii)   effects immediate or delayed;
    (iii)   known precisely by exposed persons or not known;
    (iv)   known precisely to science or not known;
    (v)   uncontrollable or controllable;
    (vi)   new or old;
    (vii)   chronic or catastrophic;
    (viii) common or dread;
    (ix)   certain not to be fatal or certain to be fatal.

Such a multifactorial approach leads to the important conclusion that
information strategies will need to be adapted depending upon the
different type of risk perception patterns of the receivers.

Work from the Robens Institute at the University of Surrey concerning stress arising from accidents and the aftermath has indicated that
sometimes counter-measures can contribute to the strain in the form of
anxiety, social disfunction and even severe depression which, at times,
may be sufficiently severe to cross the clinical threshold. It is important to
realize that humans and their society and culture cannot be considered as
objects which can be measured scientifically, just as we measure gravity
or some other scientific effect.

For ten years, Dutch researchers have studied means of introducing more school courses concerning risk into science education. Their recommendations include the use of science education in secondary schools to prepare students for coping with life in modern society and this includes teaching about ionizing radiation, which should be taken away from the nuclear physics syllabus and introduced as the topic of understanding radiation risk information. Current school teaching is criticized for focussing on closed sources of radiation, for focussing too heavily on the nucleus, for paying too little attention to risk concepts, for using a limited range of applications, and for ignoring the role of lay-ideas and lay-ways of reasoning. The point is made convincingly that, before introducing new syllabuses, the teachers have to be won over such that they themselves do not trivialize or fear the radiation concepts and risk theories that they are to teach. Industry is urged to put much effort into this co-operative teaching goal by producing course material and illustrations, *etc*. It is stressed that the best starting point is to build upon any common knowledge that lay-persons and pupils have and then to introduce additional knowledge from experts in lay-persons' language. The point is made that it is highly advisable to emphasize the difference between 'radiation' and 'radioactive material'.

There is a danger that risk could be incorrectly defined as the probability of a certain undesirable consequence, whereas, in fact, it is more rigorously defined as the expected loss of utility which can have a mathematical basis. It is often useful to view such risks from the point of view of an insurance organization. The questions that need to be asked are:

(1) Why is an insurance company willing to accept the risk which the client wishes to offload?
(2) What premium is the client willing to pay for the insurance of a certain risk?
(3) How is a less likely major loss assessed compared with the far more likely minor loss?

There are three impelling reasons as to why members of the public need to know more about radioactivity and radiation:

(1) The pragmatic reason is that people ought to be more capable of protecting themselves from the harmful effects of radiation and, also, from avoiding excessive fears.
(2) There is the democratic reason that people should be capable of reaching informed judgments in political matters involving radiation phenomena such as waste disposal, nuclear energy and exposure limits.

(3) There is the overall educational challenge that individuals often derive pleasure and fulfilment from knowing something about their environment and the world in which they live as is witnessed by the increasing number of amateur geologists and botanists in our society.

12   Risk, Perception and Social Constructions

A few years ago *The Times* Science Correspondent reported that, 'A study in 1988 showed that only 6% of Americans were scientifically literate. The rest thought DNA was a food additive, Chernobyl a ski resort and that radioactive milk could be made safe by boiling it. In short, most members of the public cannot tell a proton from a crouton.'

Not surprisingly, worries for the general population about radiation do not parallel the assessment of the statistical risks in question. They seem less concerned about increasing radiation doses due to new technologies, such as medical radiology or from radon in homes, than they do about incidents like the Chernobyl accident which generated enormous fear even though the exposure to populations outside the former Soviet Union was within the variation in mean doses of background radiation for those countries. More importantly, the Chernobyl incident increased the loss of trust in large institutions. Whereas the authorities and the experts view the risk from one point of view, the views of citizens and sensationalist press come from a different starting point. It was the 1995 Nobel Peace Prize Winner, Prof Rotblat of the Pugwash Movement, who highlighted the fact that nuclear science has moral aspects.

There is no universal definition of risk. To the lay-person, risk is often a diffuse threat, an uncomfortable possibility that something dreadful will happen. In professional terms, risk refers to a possible consequence and the probability that consequence will actually occur. Thus, if we wish to describe a situation involving risk quantitatively, we must refer to both the consequence and to the probability of its occurrence. In reality, there is often a range of consequences, with each consequence having its own characteristic probability. Thus, risk is seen to be a quantity which is multi-dimensional in nature and which cannot be unambiguously represented by a single number.

The term 'acceptable risk' is equally difficult to define. The term 'acceptable' must be judged within the context of the scenario concerned. The acceptability of a risk depends upon the associated benefits of taking a decision. If the associated benefits are zero or trivial, then the risk ought not to be accepted.

There is no such thing as a risk-free society. However, we do accept risks provided that the benefits are obvious. Thus, risks are accepted by implication. More difficult to accept are risks which arise from involuntary actions or from actions which are imposed upon the person or persons from elsewhere. The concept of unacceptable risks under normal life conditions is behind the choice of individual dose limits in radiation protection. Voluntarily, people are more willing to take much higher risks, for example, smoking, than if it were involuntary. Similarly, sometimes voluntary risk-taking seems to have an attractiveness to some people. Risk perception often relates more to the situation or the practice that causes the risk rather than to the risk itself. Risk perception is a combination of risk assessment (the identification, quantification and description of the risk concerned) plus risk evaluation (which is to do with the attitude taken by the receiver of the risk). In other terms, there is an objective risk number coupled with a subjective risk element.

When choosing numbers to describe the consequences of a risk, the International Commission on Radiological Protection (ICRP) largely deals with the probability of death. Others regard consequences in terms of the reduction in life quality. The introduction of the term 'outrage factor' is a means of semi-quantifying the public's reaction to a solution involving a risk which is imposed upon them. The opposite of the outrage factor arises from the general feeling of the existence of some benefit and also a feeling of control on a voluntary basis. Thus, trust, the ability to comprehend a situation, and the ability to control it, all lower the risk in the perception of the lay-person.

The ICRP Protection Policy is based on the assumption that, excepting accidents, doses encountered in radiation protection are not large enough to cause deterministic effects. Such effects are usually those of cell death and, in order to cause observable harm, the radiation dose must be large enough to kill a sufficient number of cells to show a clinical response. Thus, for deterministic (or quantifiable) effects, there must be a dose threshold below which the effect cannot arise and, therefore, the risk is said to be zero. Other effects, such as cancer and hereditary conditions, are called stochastic and such a dose–response relationship is assumed to show no threshold. The ICRP Protection Policy is based upon the assumption that radiation doses, other than caused by accidents, are kept so low that only the stochastic effects occur. The probability per unit dose at low doses of less than 100 mSv has to be inferred from observations at far higher doses.

The basic protection principles recommended by the ICRP for practices involving radiation are threefold, *i.e.* there has to be justification of a practice in terms of a recognized benefit, the protection involved has to be optimized, and the individual doses received must be within published risk limits.

The more that the risks are reduced, the higher the cost of further significant risk reduction and this is captured in the ALARA (As Low As Reasonably Achievable) principle of radiation protection.

There is no uniform or consistent perception of radiation risks. The public perception and, indeed, the acceptance of risk is often determined by the context in which the radiation is to be used. Also, whatever the views of the experts on radiation and its effect, it is axiomatic that members of the public view the risks differently to the technical experts. The gap between these viewpoints is to do with communication. With the exception of the publication of leaflets concerning risks from radon in the home, very few efforts have been made to improve education and communication across this gap. When a radiological disaster occurs, it is far too late to start the communication process. The envisaged pecking order concerning the risks of some 30 activities is completely different if one asks risk experts and compares their responses with (i) politically-active women, (ii) students, and (iii) a group of young business and professional citizens (Table 12.1). Nuclear power was seen to be the most risky by women voters, students, *etc.*, whereas the experts considered it far safer than other technologies. Similar lay-person surveys give a high hazard rating to terrorism, nerve gases and nuclear weapons but a low rating to sun-bathing, hair dyes and home appliances. The relative imminence of a fate has resulted in survey data showing that paralysis, male impotence, *etc.*, are feared more than a quick death.

Another interesting concept is that of equity. The risk seems to be more tolerable when the persons who bear the risk also get the benefit. In many respects, nuclear energy hazards were regarded as undesirable, in similar tone to those of chemical hazards. Psychometric studies have been used to assess a wide variety of risks and to reveal the powerful negative imagery evoked by nuclear power and radiation. In many respects, radiation shares the same dread as chemicals in that they both contaminate rather than merely damage, they pollute and befoul rather than just create wreckage and, worst of all, they create genetic damage such that the 'all-clear' is never sounded, *i.e.* the book of accounts is never closed!

To the lay-person, the use of the word 'chemical' or 'radiation' equates to dangerous, toxic, hazardous, poisonous and deadly. The expert acknowledges that some of the most toxic chemicals are, of course, prescription drugs which are very potent and toxic, but these, along with the use of medical radiation, are viewed completely differently by members of the public in that the direct benefits of exposure

Table 12.1 *The perceived risks of some* 30 *activities as envisaged by four different groups. The ranking order is from* 1 *to* 30 *representing the most to the least risky*

Activity	Experts	League of Women Voters	College students	Active Club members
Nuclear power	20	1	1	8
Motor vehicles	1	2	5	3
Handguns	4	3	2	1
Smoking	2	4	3	4
Motorcycles	6	5	6	2
Alcoholic beverages	3	6	7	5
General (private) aviation	12	7	15	11
Police work	17	8	8	7
Pesticides	8	9	4	15
Surgery	5	10	11	9
Fire fighting	18	11	10	6
Large construction	13	12	14	13
Hunting	23	13	18	10
Spray cans	26	14	13	23
Mountain climbing	29	15	22	12
Bicycles	15	16	24	14
Commercial aviation	16	17	16	18
Electric power (non-nuclear)	9	18	19	19
Swimming	10	19	30	17
Contraceptives	11	20	9	22
Skiing	30	21	25	16
X-rays	7	22	17	24
High school and college football	27	23	26	21
Railroads	19	24	23	20
Food preservatives	14	25	12	28
Food colouring	21	26	20	30
Power mowers	28	27	28	25
Prescription antibiotics	24	28	21	26
Home appliances	22	29	27	27
Vaccinations	25	30	29	29

Source: P. Slovic, in 'Radiation Risk, Risk Perception and Social Constructions', *Radiation Protection Dosimetry*, 1996, **68** (3/4), 166.

are immediately obvious. Yet another problem arises when the chemical or waste crosses boundaries and this hazard is then regarded as imported.

In perception terms, psychologists regard the 'signal value' of a risk as being the multiplication of the magnitude of the incident, times the number of people killed or injured, times the amount of property damaged times the amount of any latent effects which may well appear later. Thus, a material may well have a 'stigma', a word used by the ancient Greeks to refer to body marks or brands to expose infamy or disgrace, for example, to show that the bearer was a slave or a criminal.

Nowadays the word denotes something marked as deviant, spoiled, flawed, or generally undesirable. The presence of radiation has a stigmatizing characteristic upon our environment. Interestingly, the location of the event can have a profound influence.

In the Goiânia incident in Brazil in 1987, a caesium-137 radiotherapy source was broken into by metal scavengers and irradiated about 250 persons both externally and internally, some four persons dying. The total amount of radiation was 50.9 TBq. An Olympic stadium was used for monitoring 112 000 persons. There was a criticism that the scientific search for victims for the sake of research had created a feeling of dissatisfaction amongst those people because they felt that they had become some sort of guinea pig for the benefit of science, for which the risks of radiation are not yet well known. The incident brought a stigma to the area where hotel occupancy rates and house prices tumbled by up to 50% within the first few weeks. This ought to be contrasted with the Three Mile Island incident in the United States when no-one was injured and yet people are more aware of this incident because it was a more westernized country having optimal environmental legislation control. As with the Chernobyl disaster, it became clear that risk communication is in fact a two-way interactive process of exchange between communicators and individuals concerned.

When making risk comparisons, there is little point in getting into a panic about the risks to life unless one has compared the risks which do worry you with those that do not for your particular lifestyle. For example, an hour riding on a motorcycle is as risky as an hour of living at the age of 75. Flying 1000 miles by jet aircraft is as risky as one chest X-ray. Every single take-off or landing in a commercial airliner statistically reduces the average life expectancy by fifteen minutes. The member of the public is, of course, confused by such data. Quite simply, their attitude is, 'if it crashes – we die, if it does not crash, then we have got away with it and our life expectancy is the same'. Another figure that can be used in education is that the annual risk from living near a nuclear power plant is equivalent to the risk of being driven an extra three miles in a motor car. Such statements, however, often produce anger rather than enlightenment in the eyes of the public. The facts do not necessarily speak for themselves. When using comparative analyses, great care must be taken not to insult or deride the listener. The Chernobyl accident increased the estimated lifetime dose of 7 mSv for each person in the 2 000 000 caught up in the plume. This is equivalent to moving from New York to Denver or, in radon terms, to moving from a home in New England to one in Pennsylvania. We must be careful to standardize the terms used in our information messages and to make them clearly understandable by referring to everyday events; for example, a cup of coffee contains about sixty becquerels of radiation and a litre of sea water approximately fifteen.

It is paradoxical that indoor radon is widely-recognized as being the most important radiation burden for the general public. In some localities of countries such as Belgium, exposure levels give rise to yearly risks of one in 100. A sizeable proportion of the Belgian population lives in houses with radon levels representing a risk of one in 1000 per year (400 Bq m^{-3}). This is far higher than the total collective dose from all nuclear industries. We are talking of a significant number of deaths per year from naturally-occurring radiation. Nevertheless, the reaction of the authorities has been fairly lukewarm in terms of:

(i)   information to the general public;
(ii)   mapping of regional variations;
(iii) training of architects and building contractors;
(iv)   regulations concerning remedial actions;
(v)   quality control.

The public's acceptance of radon in the home can be contrasted with the current anxieties concerning electromagnetic fields which, as yet, have not been shown to be responsible for any deaths or damage to health.

The National Radiological Protection Board has investigated risk reduction by indirect counter-measures. These actions aim to improve the state of the affected population without significantly reducing the radiation dose received as a result of the accident. Such measures include compensation payments, improved general health care, information programmes and counter-measures to reduce other risks such as those from radon in the area.

One of the most important aspects of risk communication is that of trust. Trust in the nuclear industry, in the food irradiation industry, in the government which makes regulations and polices them, and in scientists must be worked at and built up at all times. In fact, trust is more important to conflict resolution than is risk communication. Trust is fragile, it takes many years to create but it can be destroyed in an instant by a single mishap, a mis-chosen word, or by an error in calculation. Once having lost that trust, it is almost impossible to build it back up to the former level. Abraham Lincoln wrote, 'If you once forfeit the confidence of your fellow citizens, you can never regain their respect and esteem.'

Another way of putting this is to say that negative or trust-destroying incidents are far more visible and noticeable than the positive trust-building events. Figure 12.1 shows the results of research into both trust-building and trust-decreasing incidents as assessed by 103 college students.

The social amplification of risk has demonstrated, on many occasions, that even minor incidents can produce massive ripple effects if the public

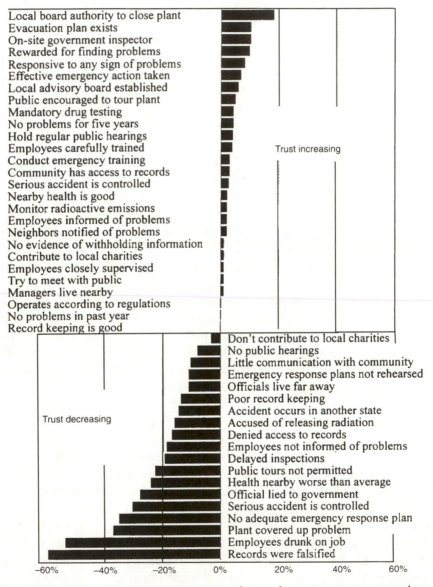

Figure 12.1 *Research into trust-increasing and trust-decreasing events assessed on a subjective scale of percentage impact upon trust*
*(Source: P. Slovic, Risk Anal., 1993, **13**, 675–682, Nuclear Technology Publishing)*

perceives them as managerial incompetence, as errors, or as other blameworthy factors.

The human factor must be considered as a safety risk. Immediately after the Estonia sank in 1994, the press published a figure that about 80% of all serious accidents are caused by human error. Whether this is a true figure or not, we must nevertheless recognize that the public expects to be protected against human error which causes accidents. The balance has changed from the days when wealth production dominated one's ambitions, often at the expense of risk reduction. Fortunately, the risk society has now reversed this relationship.

13   Risk Management

A broad classification of this subject could be into societal risks and personal risks. Much of personal risk management is involved in day-to-day decisions and will be mentioned later.

Societal risk management occurs in a community and is usually aimed at benefitting that community. The process is one of integration from numerical figures for the risk assessment through feasibility, economic implications and the societal outcomes. This last step may well involve political overtones.

At all times, there is a trend towards risk being reduced which is synonymous with saying, 'safety is improved'. However, this one-to-one reciprocal relationship is usually complicated by questions such as the viability of a suggestion, the implied cost of the suggestion and the inconvenience and disruption that will be caused whilst the new technology is being developed and installed.

Risk management may be approached from the point of view of the environment in which the worker or the dweller resides, from a past history of incidents in that area, from legislative stipulations that the activity must be as safe as is reasonably practical, of working to within well-defined legislative limits, or more recently of being under the onus of producing one's own legislative targets and working towards them.

The societal risk scenario implies that the general public is involved where this involvement can take a variety of facets. At one end of the spectrum it is a matter of them writing to the press and receiving responses through journalism. At the other end, it can be through formal depositions at public planning inquiries.

Two conclusions can be guaranteed. First, it will never be possible to reduce the risk to zero implying complete safety. Secondly, different groups within the decision-making process will surely differ in terms of the relative weighting they give to different arguments and their search for data.

The general principles of risk management may be applied from the sublime through to the ridiculous. They apply equally well to designing sets of stairs, swimming pools, planning sports activities, right through to how an elderly person dresses whilst drowsy in terms of footwear and underwear in their own bedroom. So often it is not a matter of knowing what to do in order to reduce the risk but whether we are able to do it in terms of finance, resource commitments, *etc*. The sociological pressures for a risk to be managed, and it is implied to be reduced, are exceedingly important in reaching a decision. Risk management is not an abstract activity unrelated to human thought, feeling and perception. Similarly, risk management is not separated from financial considerations. It is considerably oversimplifying the situation merely to state that 'safety is paramount' or 'all risks must be reduced to zero'. Rather, the middle ground is a combination and prioritization of risk assessment, costings, benefits and to whom concerned, political will, societal acceptance and also a knowledge of the risks of doing nothing or some alternative scheme. Throughout, it is highly desirable that public confidence is built up and maintained and that a standard approach be employed that is acceptable to the industries concerned. This formal process could well involve an estimate of the risks entailed, an evaluation of those figures in terms of personal valued judgment, and an overall conclusion in terms of the public's perception of the risk concerned. The management of these risks then becomes a skill involving communication and trust.

Writers have grouped public acceptance of risks under three headings:

(1) Occurrences of very high risk and of unacceptable consequences.
(2) Incidents of very low risk with negligible consequences.
(3) Risks which fall somewhere between these two extremes and which require manipulation by good managers in order to maximize the benefits.

It must be borne in mind that acceptance by the public may still mean that some people find the risks unacceptable.

The differences between risks assessed by experts (and they do not always agree!) and the opinions of risk-bearers can be many orders of magnitude. Further, the general public usually perceives the magnitude of a risk as far greater for high-technology industries, for chemicals and especially for nuclear areas, whereas smoking, radon in homes and gas supplies to kitchens are taken far less seriously than they merit. In addition, there is nowadays a far greater tolerance for the points-of-view of minority groupings than there was at the beginning of the industrial revolution. Thus, risks are now weighted against benefits far more than in previous times. Lee has opined that a single continuous line with a cut-off point of 'acceptability' has been replaced with the concept of 'tolerability' encompassing the idea that the risks are exchanged for some benefits.

It is easier to define what is obviously unacceptable than to define what can practically be achieved and can be accepted. The public will press for a reduction in risk which is quoted as one in 10 000 per annum and yet will be not inclined to commit their own finances to that reduction. On the other hand, a risk of one in a million per annum or better is equivalent to being killed by a prescribed drug or from a vaccination, or of an aeroplane crashing into an empty football stadium. Thus, many high-technology industries who do not make good neighbours aim for a risk level safer than one in a million per annum. This is often at considerable cost to the organization concerned.

Thus, perception can markedly influence the distribution of financial resources to tackle a perceived risk. On the other hand, public perception can work against risk reduction under certain circumstances. The cost effectiveness of implementing any risk-reducing measure has to be carefully assessed and justified before being implemented. Resources are finite. Health has to be weighed up against a healthy economy, which, in turn, has been correlated with a high standard of living, which, in turn again, has been correlated with good unemployment figures. Clearly, both the length of expected lifetime and the quality of that life must be encompassed by any final decision.

At the level of personal risk, we make many day-to-day decisions which give us a thrill or which protect us from our environmental threats. Whether to cross the road at the crossing or to risk it elsewhere is a daily decision taken by many members of the public. Whether to smoke or not to smoke, to drink or not to drink – there being a current health recommendation that we drink an average of 21 units of alcohol for men and 14 units for women per week. However, do we drink this on one occasion or do we pan it out over the seven days concerned? Do we drink it rapidly or do we drink it slowly?

Risks and threats which are involuntary to a person must be based upon a very high standard of decision-making and of public administration. Conversely, risks faced at the personal level can be more varied in terms of their threats. It is part of our culture that we get certain thrills from taking risks.

A significant part of risk management is the opposition of a community or group to a risk being built in their locality, be it a prison, an airport runway or a nuclear or radioactive waste facility. Lee has researched a moralistic model which revealed a particular aversion to radioactivity. This research revealed developments that would be opposed peacefully and within the law, but a significant percentage of those surveyed (75%) stated that they would *actively* oppose Radioactive Waste and Nuclear Power Stations in their locality, transgressing the law if need be! The lack of a resolute government policy for developments in these areas, coupled with a moralistic approach encouraged by journalistic excitement leads to active opposition.

14   Risks at Work

Every occupation involves some risk of injury or death, no matter how small and however short the working period. Often the risk is lower than if the employee had remained at home, a high risk area as just described in Section 5. The process of getting to work usually involves risks which are far higher than being at work. However, the processes of being at home and of travelling to work are voluntary, whereas the risks suffered at work are, more often than not, involuntary. The working conditions and the safety aspects have been determined by the management or by a collective group such as an employers' association to meet the requirements of the Health and Safety at Work legislation.

Many forms of employment involve the use of, or the production of, highly concentrated forms of energy. Safety measures to protect the worker from these risks can only go as far as is reasonably practicable in terms of the management's extent of compliance with the legislation by, for example, using protective guards, sound insulation, work clothing, *etc*. The risk is also dependent upon the workers' willingness to use these safety practices and upon their concentration on the job in hand.

Because of employer liability and of recent court settlements which have awarded large amounts of compensation for industrial injuries, the statistics on risks at work are fairly complete and are well-categorized. However, one must bear in mind that there is a vast range of different working conditions which affect the risk. For example, using an elderly, outdated lathe with creaky insulation and inadequate face and hand guards while standing in a pool of oil and water is thousands of times more dangerous than the automatic, computer programmed, encased lathe located in a well lit, dry and heated environment. By and large, robotics has greatly reduced the risk of work just as motorways reduced the risk of travelling. Compliance with workplace safety legislation and provision of safety features produces the same environment that is needed for the profitability and cost-effectiveness of production.

The materials used in the workplace also have an implied risk in their own right. Pit props used by miners clearly contribute to risk. The dust arising in the tailoring industry also contributes to risks to the lungs of the workers therein. Table 14.1 indicates some of the likely risks from a range of industries.

Interestingly, legislation which forbids smoking because of the use of flammable materials, or requires that nuclear workers, drivers, pilots, *etc.*, have regular medical check-ups sometimes leads to a healthy worker syndrome even though their working environment may well be of an elevated risk category compared with that of staying at home, and thus reduces the incidence of ill-health absences that may have occurred in the absence of such worker protection legislation.

Food production in the form of farming, fishing and catering has relatively high death rates. Office- and laboratory-based workers are at the other end of the risk scale.

As expected, the construction industry, using traditional methods, accounts for about ten deaths per 100 000 employees and for thousands of serious injuries. The maintenance of existing constructed buildings is even more dangerous because of falls and of objects dropped. When the construction industry is using untried technology, such as building oil rigs or river barrages for the production of electricity, there is an even greater risk of death and injury. The annual death rate amongst ocean-going fishermen can equal or surpass that of grand prix drivers. However, the perception of these by the general public is certainly not

Table 14.1   *Average annual risk of death in the UK from accidents*

Activity or Industry	Risk of death per year
Fishing	1 in 800
All accidents	1 in 2000
Coal mining	1 in 6000
Traffic accidents	1 in 8000
Construction	1 in 10 000
Metal manufacture	1 in 11 000
Working in industry	1 in 30 000
Drowning	1 in 30 000
Timber, furniture, *etc.*	1 in 34 500
All employment	1 in 43 500
Textiles	1 in 50 000
Radiation workers (1.4 mSv per year average)	1 in 57 000
Food, drink and tobacco	1 in 59 000
Poisoning	1 in 100 000
Clothing and footwear	1 in 250 000
Natural disaster	1 in 500 000
Struck by lightning	1 in 2 000 000
Nuclear accidents	1 in 10 000 000
Visiting a theme park	1 in 25 000 000

Source: Health and Safety Executive 1982.

one of equality. In terms of perception, humans tend to regard risks of death from major accidents and disasters as being far more prevalent than those from everyday workplace incidents. In fact, the truth is the opposite of this.

The National Radiological Protection Board has listed the number of days that your life expectancy is reduced as a result of various industrial hazards. It is this reduction in life expectancy rather than the cause of death that is perceived to be closer to the truth in terms of judging the threat or safeness of an industry. For example, a young deep-sea fisherman working for a year has an average reduced life expectancy of the order of 50 days per year worked, *i.e.* a person who has worked in this occupation until the age of 65 can anticipate an average of six years less of remaining life span. In the paper-printing industries, a similar calculation shows a statistically meaningless average life expectancy reduction of only ten to twelve days.

Training is an important factor in reducing occupational injuries and risks. Experience is also important. The transport of raw materials and of finished products is a major risk area and accidents involving transport vehicles account for almost half of the work-related deaths. Also high on the list are risks from sharp objects involving cuts, tears, pinches and suchlike.

One area in which these statistics are distorted is that of the chemical industry where the major risk is one of leakage and release of chemicals which are harmful by being flammable, toxic and/or corrosive, *etc.*

15   Risks from Chemicals

Approximately 75 000 chemicals are in daily use as consumer products, industrial reagents, or agricultural herbicides, pesticides or fertilisers. This number increases by about 500–1000 per year. Modern lifestyle as we know it clearly could not function in the absence of such chemicals, which are responsible for an almost doubling of our expected life at birth and an impressive increase in the quality of our healthcare, food provision, enjoyment, *etc.* Nevertheless, many of these chemicals can do serious harm in terms of teratogenesis, carcinogenesis, *etc.* Furthermore, in addition to the chemicals synthesized by the chemical industry, there are agents that are synthesized naturally by cells, such as the tetanus, diphtheria and botulinus toxins, and also aflatoxins found in infected nuts.

Fire, explosion and the release of plumes from the chemical industry have occurred from time to time, such as at Flixborough in 1984 and Bhopal which claimed 2000 lives, also in 1984.

Because energy has been used in order to synthesize a chemical, it ought not to be forgotten that it has the potential to release that energy and cause some harm. Extreme measures can be taken in order to protect humans and the environment from such harm but such risks can never be realistically reduced to zero for a chemical.

By and large, there is a threshold for each chemical whereby it can do harm to a human or to another living species. The vast majority of chemical concentrations are well below threshold levels and have a cycle through which they are introduced to a species before eventually decaying away or being eliminated by excretion. Nevertheless, the increasingly low concentrations at which analytical methods can function means that we are able to identify chemicals often well below their threshold level for harm in a wide variety of materials. For example, there are over 200 known toxic chemicals in a bottle of wine but that does not stop us enjoying many a glass of wine and apparently not suffering harm. Activists will sometimes press for the complete removal of chemicals such as pesticides, food additives and laboratory contaminants from our food and environment, but not only is this an unrealizable goal which would be a complete waste of money, but perhaps more importantly the perception of such a move would suggest that these exceedingly low concentrations of chemicals were a real threat. This would produce fear in the mind of the untrained person.

A useful example is the fact that we all contain some atoms of plutonium in our lungs and other body tissues. Some of this has arisen over millions of years but an important addition to this has been the above-ground nuclear warhead tests undertaken by the former Soviet Union, the United States, France and China in the 1960s. The professional opinion of top toxicologists in the UK is that these atoms are doing no harm whatsoever and need not be considered as a threat to our well-being. The dangers of food production chemicals creeping through into the finished product on our supermarket shelves are almost negligible compared with the threats of not having the food on shelves. Similarly, the shelves themselves are often made of composite materials glued together and so pose risks which, in theory, could pose a threat in an enclosed environment. The cleanliness of our cooking utensils and other kitchen surfaces is dependent upon the availability of disinfectants, detergents and polishes.

The toxic effects of any of these chemicals clearly depend upon the total concentrations and chemical forms (called the speciation – see later) in which they are present. A death from over-imbibing bottles of spirits, which are pleasant when drunk in small amounts, is just as dead as a death from arsenic or strychnine poisoning. In chemical toxicology, much emphasis is placed upon dose–response relationships, and

modern toxicologists now use relationships involving chemical speciation and biological response. An over simplified approach would be to
establish the dose–response relationship for all chemicals to which we
are exposed and to ensure that our actual exposure is well short of the
threshold concentrations (Figure 15.1). In practice, there is not a long
line of volunteers willing to undergo dose–response relationship
research and so data must arise from animal experiments, from tissue
cultures and from professional body epidemiology studies, such as The
Royal Society of Chemistry research mentioned later; this has been
particularly valuable to studies of workplace chemicals and cancer. The
number of cancers attributable to workplace chemicals has been
estimated as being between 2 and 30%, the exact figure being extremely
difficult to establish because of the time lag between the first exposure
to a carcinogen and the onset of symptoms. A further problem is that it
is often difficult many years hence to identify and categorize the specific
chemical and its exposure.

Another imponderable is whether an acute dose of a toxic chemical or
ionizing radiation is more, or less, dangerous than a very small dose
administered over a long time period. Indeed, in the field of radiation
protection, there is a growing number of reports of a very small amount
of radiation having beneficial effects (called hormesis), such views being
supported by increased longevity of inhabitants of regions having mildly
raised levels of background radiation.

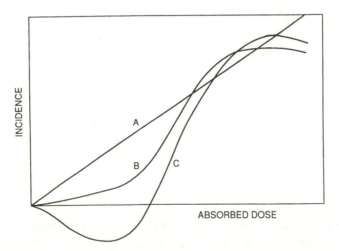

Figure 15.1   *Three types of dose–response relationship:* A = *linear;* B = *linear–quadratic;*
C = *threshold with hormesis (beneficial) effect at low doses*
(Adapted from R. L. Mlekodaj, in *Radiation and Public Perception*, eds.
J. P. Young and R. S. Yallow, American Chemical Society, Washington,
1995)

Just as one cannot define absolute safety for a chemical, so too is it difficult to achieve absolute purity. Occasionally, it is the impurity in the product which is, toxicologically speaking, more dangerous than the product itself. Clearly, chemicals introduced into humans as pharmaceuticals, foods or intravenous nutrition have to be exceedingly pure. Often, impurities are referred to in terms of parts per million or parts per billion (see the details on statistics and big numbers, Section 18). The kinetics with which the chemical reaches its site of toxicology are also important to these studies. It ought to be clear to the reader that there is no such thing as an absolutely safe chemical and that doses, means of administration and the body's ability to detoxify such chemicals (often using the liver) all have to be taken into account.

Chemists seem to be ahead of the field in terms of drawing awareness to potential risks. All chemicals are now labelled with international signs indicating whether they are flammable, explosive, toxic, *etc*. Furthermore, the use of such chemicals in the manufacture of goods, food and drinks is carefully regulated; there is a requirement for a detailed audit of the intake of chemicals into the process and the output of a finished product. Chartered chemists are made responsible under the Duty of Care legislation for disposing of unused materials, of chemical by-products and of waste to licensed companies who incinerate or entrap the chemical in a solid form for burial.

Nevertheless, there is an exceedingly wide range of chemicals to be found in use in all areas of life. They affect our lifestyle but, in turn, the risks so arising are superimposed upon those we like to regard as normal, daily life. The disposal of kitchen and garden chemicals into the drains and toilets can possibly lead to build-ups and care is taken by sewage workers lest there is a threat present from such disposal before complete dilution and dispersal has been achieved. Epidemiological work has identified clustering of certain problems because of allergies or of such build-ups.

Interestingly, many naturally-occurring chemicals are far more toxic than those manufactured by the research scientist.

Of the chemicals found in food, most of them are now listed by law as E-numbers, there being several hundred permitted ingredients as food additives under this legislation.

Sometimes, the threat is from the second or third generation of inherited materials. The primary chemical was used as crop stimulant or herbicide, which in turn was fed to animals, which in turn was fed to humans. The changes in the chemical form of materials that pass through these various life cycles can markedly affect their toxicology. It should be realized that so-called organic foods and even herbal or natural remedies all contain such chemicals. Fortunately, they are usually well below licensed concentration limits and well below the threshold levels.

Because of the broad-based nature of the field, legislators tend to err on the side of what they believe is extreme safety by insisting that levels of 'chemicals' found in food and the environment are at increasingly low concentrations – often to the lower limits of analytical detection. As the latter fall so too do the former demands for excruciatingly pure foods which can sometimes lead to the unachievable goal of a 'zero risk' product. It also raises the price of foods to a level such that some poor people can no longer afford to purchase them. One of the difficulties they face is that a lifetime of consuming the food or drink concerned could well be 80 or 90 years, whereas the additives and chemicals have only been produced over the last 10 or 15 years. Thus, real life data from complete life spans are unachievable and so the legislators err on the side of conservatism in terms of granting licences. This is especially so in terms of making the food more attractive through colours, flavours and taste stimulants, as well as the generalized area of food preservatives which enhance the shelf-life.

The workers in these industries inserting significant amounts of chemicals into the processes are, of course, protected by gloves, masks and goggles. It is thought-provoking to note that any side effects from working with large amounts of chemicals may not be the same as those of consuming much smaller amounts over a very large timescale.

In some respects, chemicals are all too readily available and can be the subject of abuse. Twice as many people die from butane gas sniffing, usually from lighter fuel bought from tobacconist shops, than die from heroin each year. A report on the BBC News on 7 November 1996 said that an average gas bottle for filling lighters will supply a smoker for approximately six years and yet a teenager who looked far too young to smoke was able to buy 34 bottles from a shop in one evening.

16   General Principles of Risk

A hazard is a set of circumstances or materials which might be the cause of harm. The likelihood of it causing harm is said to be the risk associated with that hazard, *i.e.* it is the probability that something undesired or unpleasant will happen. The nature of the undesired effect can be manifold; it may be a financial loss, injury, damage to the environment or increased worry. Having recognized the hazard and established the risk probability, the harm can be limited by risk factors. By nature, some people are risk-takers and others are risk-avoiders. Thus, risk-takers will get a thrill out of participating in dangerous sports, such as rock-climbing, but will go to great lengths to reduce the risk factors by purchasing expensive safety equipment and practising on lesser severity rocks. Clearly, opinions on the benefits of climbing rocks differ from the climber compared with those of the members of the mountain rescue team. So, too, estimates of risks to the environment caused by industrialization differ between industrialists and conservationists. Risk-takers argue that it may well be less beneficial to give up the sport and to stay at home, since their bodies will be less healthy. It is interesting that those who believe that all hazards ought to be identified and all risks reduced cannot understand why exceedingly rich leaders in dangerous sports, although rich enough to retire with plenty of spare money in hand, do not do so because they are apparently hooked on the thrill or on the admiration they receive from fans.

Clearly, there is no such thing as 'safe' and some would say that safety is the reciprocal of risk-taking. Perhaps nearer the truth is the fact that everyone wishes to minimize the incidence of undesired outcomes of taking a risk, and this is where risk-reducing factors are important. Thus, the word 'safe' is usually assumed to mean that the risks have been reduced to negligible levels.

When discussing the harm which occurs from a given risk, it is useful to have two terms in mind – mortality and morbidity. Clearly, the former refers to the risk being the cause of death and the latter refers to a non-fatal injury, discomfort and illness therefrom.

When searching for causal links between risks and outcome, it is easier in terms of mortality because death certificates are available. On the other hand, only patchy surveys are available concerning the morbidity details of hospital patients and those attending general practitioners.

It would be useful to have some indicator of harm. The International Commission on Radiological Protection has attempted to compare occupational exposure to radiation with occupational harm caused in other professions. They attempted to correlate the risk with the reduction in the quality and quantity of life, even though sweeping assumptions, such as death being equivalent to ten years' of work, were necessary. This task became even more challenging when they tried to place monetary values on the various degrees of harm. Perhaps the greatest use of such studies is to concentrate the mind and to identify risk-reducing measures that can be introduced for a reasonable price.

The strive to build new nuclear power stations in the UK has led to a thorough discussion of tolerable levels of risk. The word 'tolerable' is taken to signify the outer limits of acceptable risk. The borderline between acceptable levels for which the subject acquiesces and those levels which are just about tolerable varies from person to person depending upon their perception of the benefits that they will receive from this risk. Similarly, society as a whole has different psychological attitudes to the same risk originating from different sources. Although it may not be possible to quantify and define this borderline, it is possible to lay out general principles such that new initiatives are taken which fall only within the levels of tolerable risk.

Tolerability of risk (TOR) is illustrated in the diagram presented by Harbison, the Chief Inspector of Nuclear Installations for the UK in 1992 (Figure 16.1). Under the ALARP (As Low As Reasonably Practicable) principle, the risk will have been reduced to a certain level. TOR comes into play when one considers the expenditure necessary to reduce the risks further, and weighs up this expenditure against the benefits of such a reduction. There is an overriding implication that the costs of reducing risks further should not be disproportionate to the safety improvements gained.

Clearly, no project is commenced unless there is an implied benefit to some person or group of persons. This benefit may be financial, environmental or societal. The next stage is to assess the risk involved in establishing this project. At the lower end of the scale, there will be an extremely low level of risk which may be termed negligible in modern parlance. At the upper end, there will be the upper limit of tolerable risk and so it is assumed that the proposed project is somewhere between these two brackets. Within this broadly acceptable region, one has the chance of establishing a project provided it can be shown that the risks involved could well be negligible in comparison with other risks run by people. Assuming that there is a well-oiled network to ensure compliance

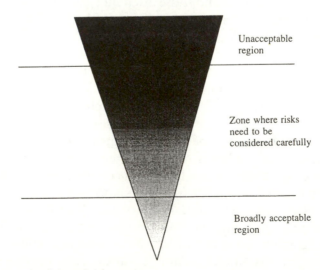

Figure 16.1   *Tolerability of risk concept*

within the regulations, the next stage is one of convincing the public and opposition that the quantification of such risks has been honest and fair.

There is a general consensus that in the non-nuclear industries a risk of death of one in 1000 per annum is experienced by workers in the riskiest of industries, such as ocean-going shipping, demolition, *etc.*

When dealing with risks and radiation, one must remember that there are several new uncertainties, not least of which is the fact that damage in the form of cancer may well occur some 10–20 years after exposure. This uncertainty leads to another order of magnitude reduction in the tolerable level. This suggests that, being industrial hazards, a TOR of one in 10 000 per annum could be assumed. Because of the image and delay of symptoms of radiation, the Hinckley Point C nuclear reactor public inquiry came to the conclusion that one in 100 000 per annum is its TOR for the general public. To put these figures into context, the annual risk of death from a traffic accident in the UK is approximately one in 10 000.

An interesting question is, what happens to the person who is surrounded by nuclear reactors? Ought there to be grounds for spreading them out equally throughout the UK, and ought we to take into consideration the fact that there are nuclear reactors just across the Channel that could well affect us? Clearly, in terms of services, maintaining expertise, workforce mobility and accumulated environmental impact, there are grounds for clustering reactors together in twos, threes or fours. Because of the uncertainties in making these calculations, it was decided at the Hinckley Point C nuclear reactor inquiry that each reactor should be considered on its own merits.

The Nuclear Installations Inspectorate's published criteria for safety assessment principles for nuclear plants are now expressed in terms of the amounts of radioisotopes released. The total predicted frequency of accidents on a plant with a potential to release into the environment more than 10 000 TBq of iodine-131 or 200 TBq of caesium-137, or amounts of other isotopes with similar consequences, ought to be less than one in 100 000 per annum.

Thus, TOR is a useful concept for defining the barriers of possible developments within the ALARP concept but can only be used provided that the ideas can be put over to, and accepted by, the public and other interested parties. It is important to have a body that can stand back and get the overview of the risk involved in a proposed development. Otherwise, each time the proposal goes before an adviser or committee there is the likelihood that another factor of ten will be added to the acceptable risk and this ratcheting of requirements will lead to non-developments in all areas of progress. On the one hand, the public and licensing bodies ought not to accept that 'nanny knows best' in terms of proposals from industry. On the other hand, it is important that the risk business does not eliminate the possibility of any development. The secret of working between these limits is one of being proactive in terms of the discussions rather than reactive to fixed limits defined by government bodies.

Fifty years after the Hiroshima and Nagasaki bombs there is still controversy about the effects of radiation and, in particular, whether there is a threshold below which radiation does no damage to a cell. The large amount of radiation that the Japanese survivors suffered has been extrapolated downwards to the very low levels of radiation experienced in normal life. This means that calculations concerning, for example, radon in homes or the effects of a nuclear accident plume are difficult to quantify in terms of expected deaths. The International Commission on Radiological Protection (ICRP) wishes to reduce the maximum level to which the public can be exposed arising from an incident from five to one milliSieverts per annum. However, the Health Physics Society in the United States opines that, below 100 mSv, statements of risk are more speculative than credible. Wilander, in his recent book *Has Radiation Protection become a Health Hazard?* suggests that it is the ready instrumental detectability of low level radiation which makes it so difficult for the public to comprehend. He states that, if Chernobyl had released a chemical rather than radiation with similar toxicology, it would have been almost impossible to detect more than ten or twenty miles away from the plant, and in distant Britain and Scandinavia no-one would have been worried. The art of radiation measurement is so advanced that the plume could be detected and it has been calculated that, statistically, Chernobyl is expected to cause 25 000–30 000 deaths over the next fifty years if one assumes a linear, no-threshold dose–response curve (Figure 15.1). But, this is a very small figure compared with the 700–800 million who will die of cancer during that time period from a population of 4.3 billion living in the northern hemisphere. In fact, it is about three- to four-thousandths of one percent. The UK National Radiological Protection Board (NRPB) has examined the Hiroshima figures and is still of the opinion that there is no threshold below which radiation can be assumed to do no harm. At really low levels, the risk may be insignificant compared with normal, everyday risks but, nevertheless, it is not to be assumed to be zero. This fail-safe viewpoint called ALARA (keeping doses As Low As Reasonably Achievable) is taken because the energy released by some forms of radiation, when they cut across both strands of DNA and RNA genetic codes, possibly initiating cancers, is far more harmful than similar types of energy from other sources.

Counter-arguments propose that at very low doses there are adaptive responses, biopositive effects (called hormesis), and there has been criticism of the ICRP and NRPB for interpreting 'reasonably' in ALARA as 'as low as possible, at any cost'. They quote the wastage post-Chernobyl of destroying good food, inducing abortions and psychological and economic damage from the Chernobyl plume.

17 Defining 'Safe' and 'Safe Enough'

By now, the reader will have realized that defining the word 'safe' as a zero risk is not practicable and can never be attained. Rather, the word 'safe' can be taken to mean 'safe enough' and possibly 'even safer'. Very rarely is it feasible to reduce the risk and to maintain all the benefits. The first and obvious example of this is the fact that all moves to reduce the risk involve some cost and, therefore, product price-competitiveness comes into question. The second example is the probability of contracting cancer from chemicals contained in our water supply. This stands at one in 1 000 000 over 50 years of drinking. Some would argue that public water is never to be drunk again but even they would realize that it is just not feasible for everyone to be drinking highly purified, doubly-distilled or de-ionized water to which the essential traces of salts (of limited purity!) have been added. Realizing that society as we know it today would grind to a shuddering halt if some of our colleagues were never exposed, occasionally, to increased risks, then it is clear that we have to go for measures that protect society as a whole throughout their expected lifetime. The crisis in larger numbers referred to in the drinking water example above ought really to be seen in terms of the public's confidence in the scientific, professional and societal management of water supplies and other features of our civilization.

Having reached a reasonably happy medium in the plot of risk *versus* cost, it must be realized that attempting to make an industry extremely risk free, and not vulnerable to threat, could well increase the risk. Two examples will illustrate this point:

(1) It is well known that a vast proportion of the World's gold supply is locked away in Fort Knox, USA. The concentrating of such wealth in one location very much attracts the interests of would-be robbers and increases the amount of risk to those guarding the emplacement.

(2) Taking the example of risks of carcinogenesis arising from drinking water, it could well be argued that workers in the water industry tending reservoirs, rivers and chemical treatment plants ought to be very well protected in terms of face masks, waterproof

footwear, rubber gloves and so on. The injury from accidents arising from slipping and from the discomfort of wearing these protectives would vastly exceed any risks to which they are subjected because they are working with the water supply and its chemical treatment. It clearly would be a ridiculous situation were we to envelop employees in every piece of imaginable safety equipment. It is thought-provoking to note that, although they are protected from chemicals when concentrated, we do not wear protection when we drink tap water. Yet, analogous to radiation and a linear dose–response relationship, there could be statistical deaths associated with imbibing the chemicals present in dilute form in drinking water!

The UK Department of Health has recently addressed the topic of the language of risk and definition of safety. They emphasize that the language of risk often misleads rather than informs. Only when the language is correct can one legitimately compare one risk with another. Particular thought must be given to which factors convert a hazard into a risk, and it must also be realized that, although we can now estimate the probability (say of one in 1000) of an adverse effect influencing a population at risk, we can never pick out the one in 1000 persons who will be so affected. Retrospectively, of course, it may be possible to identify those who succumbed to this risk, after any other factors have been taken into account. The Department of Health recommends a Hazard Analysis Critical Control Point (HACCP) process.

In general, whenever a patient has an adverse reaction to a treatment, therapy or surgical procedure it does not imply that there has been negligence. Rather, it reflects the inherent risk in the procedure or therapy. This may occur because of the incidence of false positive or false negative tests at the biochemical or pathological laboratories; it may be because different patients react differently to the same dose of drugs; it may be because of patient non-compliance or perhaps through lack of immunity or repair deficiency syndrome in the individual. What is important, and is the direct responsibility of the clinician, is to watch out for these warning signs and to monitor the patient throughout so that antidotes can be given to reduce the undesirable reaction.

The risk of venous thrombosis from using combined oral contraceptives is far lower than the risks of pregnancy and childbirth. The Department of Health definitions of terms such as negligible risk, moderate risk, unknown risk, *etc.*, are given in Table 17.1, and distinguish between mortality and morbidity. It is interesting that they suggest that, if the word 'safe' is used, it must be seen as meaning 'negligible' but ought not to be used to indicate 'zero' risk. It is important to realize that

Table 17.1   *UK Department of Health comparative risk table*

Event	Risk
Transmission to susceptible household contacts of measles and chickenpox – adverse effect	1 in 100 = High
Transmission of HIV from mother to child (Europe) – adverse effect	1 in 100 = High
Gastro-intestinal effects of antibiotics – adverse effect	1 in 100 = High
Smoking ten cigarettes per day – death All natural causes, age 40 years – death	1 in 100 to 1 in 1000 = Moderate
All kinds of violence and poisoning – death Influenza – death Accident on road – death	1 in 1000 to 1 in 10 000 = Low
Leukaemia – death Playing soccer – death Accident at home – death Accident at work – death Homicide – death	1 in 10 000 to 1 in 100 000 = Very low
Accident on railway – death Vaccination-associated polio – adverse effect	1 in 100 000 to 1 in 1 000 000 = Minimal
Hit by lightning – death	1 in 10 000 000
Release of radiation by nuclear power station – death	= Negligible*

*1 in 1 000 000 = Negligible = 'safe'
Source: Department of Health Annual Report for 1995.

procedures meriting these titles can be modified and improved from time to time which means that they can be described with a more optimistic phrase. Such steps as reducing the side effects and being aware of which parameters to monitor clearly contribute to this progression in a positive direction on the safety scale.

18   Statistics – Dealing with Large Numbers

Although perception and the relative importance placed upon an event by the general public are definitely not quantitative, everyone realizes that the data and statistics concerning the incidence of events have to be as reliable as possible. The collection of such data is subject to many difficulties. The morbid patterns of illness, injury and stress caused by an event are difficult to assess in a quantitative manner. Even recording the number of deaths is insufficient information in itself, as the main criteria which determine the perception of the risk of that event are those of premature death and a lessening in the quality of life. Nevertheless, we must start somewhere with data collection.

By and large, the task of risk quantification falls into three broad categories:

(1) Risks described by a plentiful supply of statistics and for which the data have been accurately recorded.
(2) Risks for which data concerning the hazard and also data concerning the incidence of the harmful event have both been carefully recorded, but for which the linking relationship, which may well be causal, is difficult to establish. An obvious example of this is parental exposure to radiation and the development of childhood cancers in offspring some decades later.
(3) Risks of harmful occurrences which have yet to happen and which need to be based upon forecasts of probabilities some time in the future.

The BMA divides risk assessment into two areas. The first is *risk estimation*, which is very much a scientific activity based upon scientific judgment of whether the findings are statistically significant. For example, statistically there may, or may not, be clusters of the condition at a location.

The second aspect of risk assessment is that of *risk evaluation* which is a very much more subjective topic involving sociological, political and perceptional aspects. Table 18.1 indicates risks of a high probability and of a large magnitude which are accepted with a certain degree of complacency by Earth dwellers today. On the other hand, because risks arise from the new, somewhat feared, technologies and are possibly encouraged by the style of press reporting of these risks, the perception of a low-to-moderate risk may be far higher than that held by technical experts (Table 18.1).

Although counting the number of deaths has been a means of collecting data for a century and a half, obtaining further information concerning the social class and previous history of the unfortunate deceased is obviously difficult. This means that establishing causal links with the exact source of the harm is often more difficult than one would imagine. For acute deaths arising from an obvious accident it may be possible to categorize the cause of death, but if one tries to dig deeper and establish the blood alcohol level of the person involved such information will clearly be difficult to acquire as it may not be released until after a court has deliberated some year or so later. Similarly, the dietary,

Table 18.1 *Perception of the public compared with that of experts*

	Risks as perceived by	
	Experts	*Public*
Nuclear power and waste	A moderate risk which is acceptable	An extreme risk which is unacceptable
Radon in homes	A moderate risk which merits positive action	A very low risk met with apathy
Irradiation of food	A low risk which is acceptable	A moderate to high risk whose acceptability is open to question
Electric and magnetic fields	A low risk which is acceptable	A risk which is generating increasing concern and becoming unacceptable

Source: P. Slovic, *Radiation Protection Dosimetry*, 1996, **68**, 165–180.

smoking and work exposure habits of the deceased are also difficult to assess with any degree of reliability. Nevertheless, certain learned professions have established databanks stretching back over many years concerning the lifestyle, employment characteristics and exposure to certain chemicals for their members. One such catalogue is that compiled by The Royal Society of Chemistry, and over the years they have been able to publish causes of death for 4012 out of 14 884 deceased members followed for 25 years alongside a list of the chemicals to which they were exposed at work.

There are many paradoxes which arise from this data collection. Accident and emergency departments of hospitals reported treating far worse injuries since the introduction of seat belts. Further investigation showed that there were fewer of them, but it became clear that these seriously injured persons who had been wearing seat belts would have been dead on admission in previous times before their use. Similarly, as trauma units are now available from accident and emergency ambulances there has been a swing in these statistics towards admitting more very seriously injured and ill patients with a less prevalent death rate.

A real, thorny problem surrounds the response between dose (of radiation, poison or energy such as the Sun) and the extent of the injury. We have seen that dose–response curves may be linear, or bent, and they may have a threshold level below which there is no noticeable injury and which is not accumulative (Figure 15.1).

Another difficulty arises when one wishes to describe numerically quantified risks in terms of words understandable to the lay-person. It ought to be a golden rule that risks are never quoted without reference to a standard equivalent group elsewhere from the suspected source of hazard. Furthermore, it is important to define openly the size of sample; for example, is it three deaths from a boating accident per family, per yachting club, per 1000 yachtpersons, or per population of Britain?

One of the greatest uncertainties lies in being able to make future predictions of the likely incidence of harm arising from a quantifiable risk. By and large, such estimates tend to overestimate the likely incidence slightly, but in terms of assessing the hazard this is equivalent to erring on the safe side.

Table 18.2 lists the average risks of death per person per year for UK dwellers. There is ambivalence as to whether one records risk as 'one in a number', or 'one in 100 000' (which tends to be used by the UK Department of Health) or as a percentage. The figures given in the tables in this book use all of these. In all instances, it is important to state

Table 18.2   *Risk of death per person in the UK per year. Level at which UK population would commit own resources is approximately 1 in 10 000*

Smoking 10 cigarettes a day	1 in 200
Heart disease	1 in 300
Cancer	1 in 400
All natural causes, age 40	1 in 850
Any kind of violence or poisoning	1 in 3300
Influenza	1 in 5000
All accidents	1 in 3000
Accident on the road	1 in 8000
Leukaemia	1 in 12 500
Playing soccer	1 in 25 000
Accident at home	1 in 26 000
Accident at work	1 in 43 500
Natural background radiation	1 in 40 000
UK nuclear industry	1 in 40 000 000
Radiation working in radiation industry	1 in 57 000
Homicide	1 in 100 000
Accident on railway	1 in 500 000
Hit by lightning	1 in 10 000 000
Release of radiation from nearby nuclear power station	1 in 10 000 000
Suicide	1 in 13 500
Murder	1 in 160 000
Accidental poisoning	1 in 170 000
Theme park visitors	1 in 25 000 000
Travelling to visit theme park	1 in 3 500 000
Hangliding	1 in 1500
Air travel per kilometre flown	1 in 2 500 000 000*

*8 × distance to Sun and return or 60 000 orbits of the Earth
Source: OPCS, 1986.

the time interval being assessed. Venturesome statisticians have tried to define a Richter-type scale of Safety Degree Units but this seems to have fallen out of fashion because of the difficulty of communicating it to the general public. This is not surprising since, if one is aware that one's home has been built to withstand an earthquake up to scale 8, that is the sole number one is interested in knowing at the time of a tremor. On the other hand, if the likely risk of an event in safety degree units is 8 (that is, a risk of one in one hundred million), one's *perception* of that hazard is still dependent upon many factors other than the cold SDU number.

As was shown when investigations were made into alleged clusters around Sellafield and Dounreay, by and large, historical data are usually pretty useless in terms of being able to quantify a risk and predict the future. Similarly, animal experiments using small rodents have very limited use for us in terms of estimating future likely damage to humans. Finally, high dose testing of a potentially hazardous material will not necessarily indicate the consequences of a rather low dose exposure over a long timescale. Animal experiments are useful in indicating the No adverse Effects Level (NEL). This is often expressed in terms of mg kg^{-1} of body weight in order to extrapolate to humans, but some pharmacologists would argue that it is more reliable to do it in terms of the surface area of skin rather than per kilogram of mass. Other figures refer to acceptable daily intakes using similar units. TLVs, or Threshold Limit Values, are placed upon workers in the chemical industry. Bearing in mind that more than one thousand new chemicals are produced each year and that it can cost £½m to establish the likely toxicity of a chemical, the expense of being the producer of new chemicals is exceedingly large. The BMA has pointed out that, were salt or sugar to be tested as potential food additives using modern standards, it is highly unlikely that either would be permitted for use in our food!

The likely risks of construction and engineering are also difficult to assess. The use of fault trees or event trees is widespread but critics claim that they underestimate the effects of human error.

Estimates of likely risks can vary exceedingly widely. A study in the United States which examined the likely expected increase in the number of human bladder cancers in the USA from 70 years of exposure to an intake of 120 mg of saccharin per day varied between 0.22 and 1 144 000 additional deaths!

From the UK register of fatal accidents, it is possible to estimate the death rate on UK registered airliners from crashes. The figure for the 1980s is 0.04 deaths per one hundred million passenger kilometres. This is a figure of one death for every 2400 million passenger miles. These exceedingly large numbers are difficult to comprehend alongside the headlines of an aeroplane crashing on take-off or landing. Others would argue that the figures ought best to be expressed in terms of the number of sectors (*i.e.* one take-off and landing) flown by the persons concerned.

It is important to realize that steps taken to control or minimize a recognized risk may sometimes present risks themselves. The British Medical Association Board of Science and Education has highlighted this in respect of their views concerning the irradiation of food with X-rays or gamma-rays in order to inactivate micro-organisms or parasites. They draw the reader's attention to the fact that the irradiation reduces the levels of some vitamins, specifically vitamin C and thiamin, and that such low doses of radiation would have no effect upon viruses, enzymes or toxins already contained within the food.

When things go wrong involving energy or the chemical industry, they tend to do so on a scale which writes its own headlines. A liquid petroleum gas storage depot in Mexico City exploded in 1984 killing more than 500 people. Figures estimating the risk from the UK's largest petroleum storage centre at Canvey Island suggest there is a one in 100 per year chance of an accident killing ten people, and a one in 5000 chance of an accident killing 10 000 people! These calculations tend to be based on a worst possible case scenario, and in the event of an incident occurring the death rate would probably be lower. The Chernobyl nuclear reactor explosion caused more than 30 deaths and approximately 200–300 received medical attention.

The reports on Chernobyl suggest that 135 000 people were evacuated from within a radius of 30 km of the plant. In excess of 200 people received doses of over 4 Sv which, as a rule of thumb, will kill 50% within 60 days. Assuming a linearity of dose–response relationships, the ICRP figure indicated that approximately 100–125 deaths will occur for every 10 mSv per million persons irradiated. This permitted them to calculate the expected extra cancer deaths in Western Europe arising from the Chernobyl plume, 37 of which would occur in the UK (Figure 18.1). A figure of 1250 extra cancer deaths would occur in Eastern Europe and 7500 extra fatal cancers in the former Soviet Union over the next 50 years. However, this begs the question as to whether exceedingly low doses over a very large number of people can be reliably stated to give a statistical number of expected deaths. This is especially important when one realizes that, in the UK, approximately 140 000–150 000 people die from cancer each year or 7 000 000 over the 50 year timespan of the Chernobyl figures, and that the extra 37 deaths will not be evident amongst the usual fluctuation of ±6000 deaths from cancer per year. The dose throughout Europe from Chernobyl was less than one-fifth of the normal background dose. In England it would be equivalent to the dose received in flying to Spain and back in an aeroplane, or to taking a three-week holiday in Cornwall where there is an elevated background

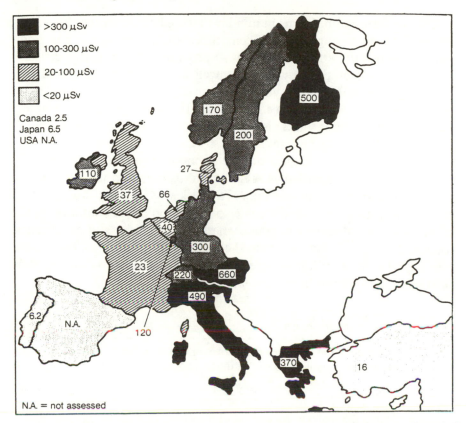

Figure 18.1 *Additional statistical cancer deaths computed as arising from Chernobyl Incident* (1986) *emissions over the 50 years from 1986–2036*
(Adapted from *The Radiological Impact of the Chernobyl Accident in OECD countries*, OECD, Paris, 1987, ISBN 92-64-13043-8-F 140)

radiation level compared with the rest of the UK. As stated elsewhere, the radioactivity limit for caesium-137 in sheep and lamb meat consumption in Cumbria and Wales (heavy rainfall and the plume doubling-back caused six times the UK average precipitation) was set at a level lower than that of many foods routinely swallowed, such as nuts and instant coffee granules. Clearly, the Chernobyl analysis has underlined the immense difficulties of keeping statistical deaths and injuries in perspective with respect to everyday occurrences.

One of the increasing difficulties in dealing with these very large numbers is that we are now able to analyse much lower concentrations. Whereas one part per million used to be the accepted limit for analytical chemistry, there are now many techniques which get down below levels of one part per billion. This is equivalent to a grain or two of salt in a large swimming pool, or the thickness of a piece of plastic compared with a 1000 km journey.

Another figure to be borne in mind is that we have to accept life as it is rather than as it might have been. Since radiation and other high-technology solutions to problems were provided earlier in this century, we have to accept the downside as well as the upside of these life-extending and quality-improving breakthroughs. By and large, we swap one problem in life for another. Hopefully, the newly presented problem will be less of a threat and risk than the one that is being treated or handled by the new technology.

When one turns to natural disasters, one can immediately envisage that they are often regarded as acts of God (as in insurance agreements) and as being more acceptable than if the risk was imposed upon one by industry. Approximately one quarter of a million people per annum are killed by natural disasters such as tornadoes, landslides and avalanches, volcanoes, *etc*. One earthquake in the Peoples' Republic of China killed over a quarter of a million people in a single event. Approximately 1800 persons are killed on average per earthquake.

Table 18.3 *Large numbers which the public appear to accept with complacency*

5 500 000 000 persons are registered as the latest World population figure (1991).

58 606 000 was the UK population in 1995 and is expected to rise to 61 million by year 2020.

14 million of the UK population (*i.e.* 25%) are dependent upon a Department of Social Security giro cheque.

15 million of the UK population are aged 60 or over.

800 million of the World population are malnourished and hungry.

175 000 of the UK population die each year of cardio-vascular disease.

140 000 of the UK population die each year of cancers.

250 000 of the World population die each year in road traffic accidents.

6000 suicides occur each year in the UK.

20 000 UK deaths are attributable to hospital acquired bacterial infection.

300 000 of the World population are bitten by snakes and 40 000 die each year.

3100 deaths, 35 000 injuries and 70 000 near-misses have occurred over the last 20 years of hostilities in Northern Ireland.

3500 chip-pan accidents occur per annum and ∼20 deaths occur in the UK.

10 000 of the UK population die each year from conditions arising from fume particles from engine exhausts (particles less than 10 μm diameter). Cardiff has the highest pollution compared with Birmingham, Newcastle and London.

650 000 births each year in UK cost approximately £1700 each to the National Health Service.

645 000 of the UK population die each year.

70 000 of the UK population die each year from smoking-related conditions. For every 1000 young male adults, one will be murdered, 6 killed in road accidents and 250 will die before expected lifespan because of tobacco smoking.

50 000 000 working days each year are lost because of smoking at a cost of £500 000 000.

£19 000 per minute is added to the UK Exchequer from tobacco taxes.

300 non-smokers in the UK population die each year from secondary tobacco smoking (passive smoking).

Continued

The human mind seems to be able to accept and block out risks arising from natural disasters. The Californian earthquake of October 1989 killed over 270 people. There are almost daily earthquakes in California. There is an estimate that there is an almost 50/50 chance that a catastrophic earthquake equivalent to that of 1906, which flattened the whole of San Francisco, will occur within the next 30 years; this equates to between a one in 20 and one in 50 chance that it will happen during any given year. Nevertheless, these threats are not uppermost in the minds of Californian dwellers, and the populations of these cities continue to expand. At the other end of the extreme is the worry that a parent has when a child visits a theme park. The UK Health and Safety Executive has indicated that there is a one in 25 000 000 chance of a fatal accident per annum. Nevertheless, it is natural for parents to worry. Because of a lack of education in risk perception and communication, we accept many large numbers with complacency (see Table 18.3) and yet fear others.

Table 18.3   *Continued*

£2 000 000 000 is spent on the UK population each year treating alcohol mis-use, alcohol being the most potent psychotoxic drug legally available.

$US 8 000 000 000 is spent in the USA each year dieting to treat obesity. The World's heaviest man weighed 100 stone (J.B. Minnoch, 1983, in USA) and in the UK, P. Yarnall weighed 58 stone when he died in 1984 aged 34.

30 000 lives lost per year in the USA are claimed to be caused by obesity.

1000 airline passengers die in flight each year out of 1 000 000 000 flights and 4000 die soon after landing.

250 000 of World population die each year of natural disasters.

32 330 000 hold driving licences in the UK, of whom:
    1 954 000 are over 70 years of age, of whom:
    95 000 are over 80 years of age, of whom 3 are over 100.

5 000 000 criminal offences are recorded in England and Wales each year.

£24 000 is the average cost of keeping a prisoner in a UK jail per year.

170 000 000 000 megawatts of solar energy each year reach Earth.

1 000 000 tonnes of radioactive low level waste (LLW) will have accumulated by the year 2000, plus:
    160,000 tonnes of ILW in total to year 2000, plus:
    4000 tonnes of HLW in total to year 2000.

5–10 000 000 tonnes of hazardous industrial wastes are produced each year.

50 000 000 tonnes of coal mining spoil are produced each year.

20 000 000 tonnes of refuse (England and Wales) are produced each year.

155 000 Bq m^{-3} radon levels are recorded in some Peak District caves compared with 200 Bq m^{-3} action level for homes.

44 000 miles is the distance traversed by data from every National Lottery ticket via Satellite to the Watford Central Centre.

Water costs less than 0.1p per litre but purchased as cleansing water as a cosmetic costs up to £145 per litre.

1 g of plutonium produces energy equivalent to 100 g of uranium or 1 000 000 g of oil.

7 200 000 000 penny coins are in circulation.

An interesting new angle on natural disasters arises from compensa-
tion-seeking litigation. Lawyers pursue those in authority who failed to
predict, and to prepare against, floods, earthquakes, avalanches, *etc*.
They would have us believe that compensation for every misfortune is
every citizen's right!

19   Radioactive Waste Disposal

We are now living in the sixth decade of the nuclear age, but during the
lifetime of many of us still researching in the field humans have gone
from 'pony' through to 'plutonium' power, *i.e.* much of the energy
required for our lifestyle at the beginning of this century was horsepower,
whereas now a substantial and important fraction comes from nuclear
power. This presents a vast field in which many of the principles of
dealing with large numbers described in the previous sections are amply
illustrated, since decisions are being taken in terms of billions of
Becquerels of radioactivity (a Becquerel is a disintegration per second)
and up to a million years into the future in terms of waste disposal. Thus,
this section discusses the new scientific problems and some solutions
which have arisen with the use of nuclear energy and the disposal of the
wastes therefrom. It also discusses the question of how one objectively
and scientifically composes an environmental impact assessment for the
disposal of any wastes which are radioactive and/or which contain
chemicals.

This planet, based upon approximately 100 elements, was established
around 4500 million years ago when a blinding supernoval explosion in
outer space gave us the solar system, when, under the impact of gravity,
the heaviest elements were compounded into the molten core, the
middleweight elements (such as transition series metals) were in the
mantle, and the lightweight elements at the top of the periodic table were
gravitationally placed into the crust. Bearing in mind that four-fifths of
the Earth's surface is covered in water, one can see why humans, who
evolved from primitive cells on the beaches being washed with sea water
and bathed in sunlight (the Earth lacked its protective atmosphere of
ozone in those days), are closely related to the composition of ancient
seas and beaches, and are also dependent on essential elements which
appear at the top of the periodic table. Vestiges of the first transition
series elements are essential to metalloenzymes involved in enzyme
activity and gas transport, but until humans started to mine deep down
into the mantle, precious few of the other elements were found in our
environment and in our bodies. (There is a notable exception to this
generalization in the form of uranium-based nuclear reactors which were

established when natural materials exceeded critical masses millions of years ago and which have long since become extinct. The elemental plumes from these extinct reactors can be found in the plant life near to their locations.)

Thus, we have a lightweight elemental body composition being exposed a few thousand years ago to the heavy precious metals such as gold, silver, platinum, *etc.*, and then being further challenged with the industrial revolution and mining of metals some two centuries ago; an even further challenge appeared from 1939 onwards when Curie, Savitch, Hahn, Strassman and Meitner introduced us to nuclear fission. Thus, a range of fission products, as well as heavier elements of the periodic table, are now found in humans and, for example, since the above-ground nuclear warhead tests during the 1960s, particles of plutonium can be found in the lungs of all animals (including humans) on this planet.

Nuclear fission is dependent upon the stability of the nuclei of large atoms. Atomic nuclei possess exceedingly large concentrations of energy because of the co-location of positively charged protons within a very small volume indeed. Typically, the radius of a nucleus is of the order of 10^{-14} m. When the nuclear forces which hold this energy intense particle together exceed breaking strain, the nucleus will shatter and give rise to fission products very roughly ⅓–½ the mass of the original nucleus plus a whole variety of smaller fragments. The usual incident which can trigger off such fission (this is akin to adding the last straw to the camel's back) is one of bombarding a nucleus with a neutron (Figure 19.1). The addition of this extra particle exceeds the nuclear forces that hold the nucleus

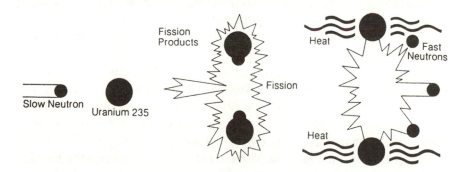

Figure 19.1 *Nuclear fission in which a slow moving neutron strikes the nucleus of a uranium-235 atom causing it to split into two pieces which fly apart generating heat. At the same time neutrons are given off which, when slowed down by the moderator, are able to split other uranium-235 atoms, thus maintaining a chain reaction by splitting atoms and a steady generation of heat*
(*Source*: UK Nirex Limited)

together and the whole core fissions or shatters. Thus, a nuclear reactor works by the neutrons produced from one fission being used to trigger off the fission of another nucleus.

The 1940s saw a race for producing nuclear warheads, and then in the 1950s the emphasis was changed to that of producing nuclear power. However, the penalty that had to be paid for this so-called progress was that a wide range of fission products was produced as waste. Each one has its own particular chemistry and its own particular radioactive properties such as radioactive decay characteristics and the type of particles emitted. Nowadays, the greatest challenges are the safe disposal of this radioactive waste and the decommissioning of the early genera-tion reactors.

Lest this article over-emphasizes the difficulties of the disposal of radioactive waste arising from nuclear fission, let it be stressed that there are manifold benefits which can be credited to our introduction to nuclear fission earlier this century. For example, World peace seems to have been maintained, more or less, by the threat of a nuclear holocaust; a plentiful supply of nuclear energy has enabled some countries to surge ahead in terms of their competitiveness when they did not have indigenous sources of energy within their boundaries; many of us are alive today because of the immense contributions that nuclear medicine and radiology have made in the areas of medical diagnostics and therapy; and multiple uses of radiation in the high-technology industries would not have been possible without nuclear fission.

Turning to the challenge of waste disposal, it must be remembered (i) that radioactive waste is a mere proportion of the sum of the many wastes that we produce, and (ii) that for every element of the periodic table there will be a cycle in nature equivalent to the carbon and nitrogen cycles that we learned in our school biology days (Figure 19.2). Many of the newer elements are actually radioactive isotopes of existing biologi-cally-necessary elements and so pass through exactly the same cycles. This is why, for example, radioactive sodium-24 is used for plotting blood flow abnormalities in humans since the radiation it emits can be detected from outside the body, allowing blood supply restrictions and clots to be located. Other elements that are not naturally-occurring develop their own cycles whereby they find partners in the soil and groundwater, and then get taken up into plants, animals and the humans that devour these species.

To summarize so far, new elements have been introduced relatively recently into our evolution and, although the principle of the survival of the fittest still prevails as proposed by Darwin, we have not had time to adapt to their presence. The solutions to this environmental challenge to our quality of life can be one of either ceasing the technology – but then

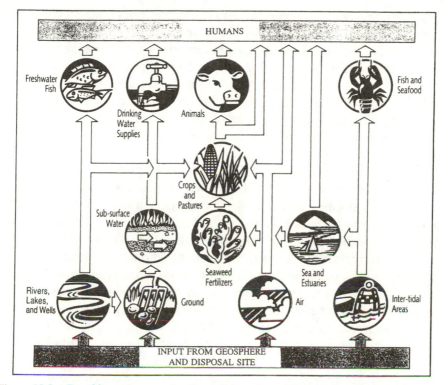

Figure 19.2 *Possible pathways of radionuclides released from a repository in the geosphere and travelling to humans*
(*Source*: UK Nirex Limited)

we are still left with the backlog of new elements and radioactive elements awaiting disposal – or alternatively of minimizing the threat to humans through the safe handling of the new technology and the safe disposal of these new radioactive isotopes.

One fact is clear. The problem will not go away. The accumulated wastes, some of which are now more than 50 years old, will need to be stored and disposed of safely and in a professional manner. It must be realized that nature has offered us 340 isotopes, some 70 of which are radioactive. Then came the nuclear age and the ability to synthesize elements towards the end of the periodic table, which effectively released a further 900 radioactive isotopes. Each of these will have a different radioactive half-life and may give out alpha, beta or gamma rays. Under the wrong circumstances, such radiation can lead to cancer or to genetic defects. It has been estimated that some 6 000 000 g of plutonium have been liberated into the atmosphere since the beginning of the nuclear age. It must be stressed, however, that the increased threat of radiation causing diseases is not in the ratio of the 900 radioactive isotopes produced this century *versus* the 70 radioactive ones which have always been present in nature. The threat, of course, depends upon the amount of radiation, the location on to which it imposes (into bone marrow or on to the surface of the skin, *etc.*) and to the type of radiation and its energy. Thus, there are many roles in which science can contribute in terms of tackling the safety aspects of this nuclear era in which we live. Certainly, becoming non-nuclear and turning our back on the problem is not a good way forward since we now have a considerable build-up of material to be dealt with safely.

Different countries have a different reliance upon nuclear power as a source of their energy. In the western world, provided that they are not residing in extremes of climate, an average person requires approximately 3 kW (a three-bar electric fire) of energy to be switched on at the moment of birth and to be extinguished when they die. This is the energy required for their heating, transport, house and office construction, food, clothing, *etc.* Developing countries require somewhat less but are rapidly catching up with the developed world in terms of their energy requirements. Wise governments choose a portfolio of energy sources involving natural sources, fossil fuel-based sources, renewables, nuclear, *etc.* This means that they are not totally dependent upon the supply of raw materials from one source. An interesting reflection of this is the cost of uranium which soared in the 1970s but then came down again when it was decided to construct the Thermal Oxide Reprocessing Plant at Windscale to recycle uranium (Figure 19.3).

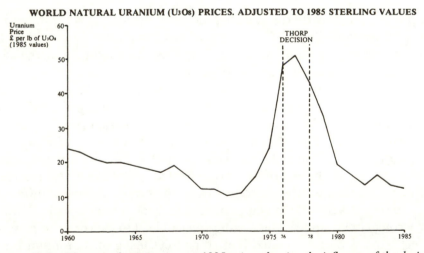

WORLD NATURAL URANIUM (U₃O₈) PRICES. ADJUSTED TO 1985 STERLING VALUES

Figure 19.3   *The cost of uranium ores at 1985 prices showing the influence of the decision to construct THORP (Thermal Oxide Reprocessing Plant)*
(Reproduced with permission from HMSO, First Report from the Environment Committee, vol 1, 1985–1986, ISBN 010-280586-5)

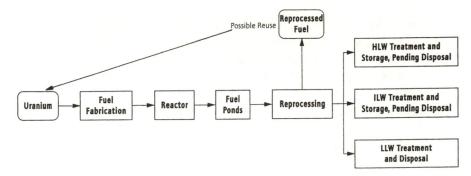

Figure 19.4   *The nuclear fuel cycle including reprocessing and the three categories of waste produced*
(*Source*: Nuclear Electric Limited)

Concentrating on uranium as a source of energy, wastes are produced at all stages of the energy cycle commencing with the mining and then moving on to fuel enrichment, the machining and canning of the fuel, its recycling, *etc.* (Figure 19.4). The fact that this technology produces waste ought not to be a surprise since all forms of energy produce waste in one form or another (Scheme 19.1). We can recall the large spoil tips arising

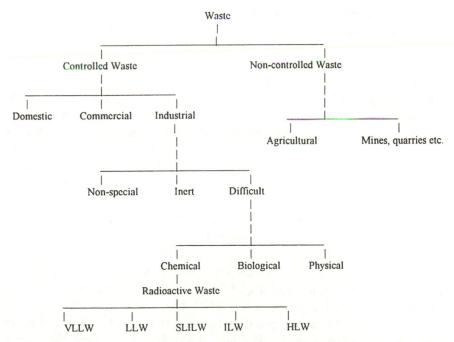

Scheme 19.1   *The waste clarification system used in the UK; approximately 140 million tonnes of controlled waste are produced each year and some 3000–4000 million tonnes of non-controlled waste. All wastes in the 'difficult' subgroup require a licence for disposal*

from coal mines, the plume from oil production when leaks occurred, the acid rain from fossil fuel-burning stations, and in the 19th century there are numerous reports of grass growing in St. Paul's Cathedral in London because of the waste from horse-power being carried in on the boots of worshippers!

Some of these wastes are present in amounts similar to the background level, for example, the soil from a nuclear site or tissues used in wiping surfaces therein; others are found at exceedingly high levels of up to 10^{12} MBq m^{-3} (the unit is a megaBecquerel per cubic metre, *i.e.* a million disintegrations per second occurring in each cubic metre of sample). This last figure refers to the activity from spent nuclear fuel rods as they are taken from a nuclear reactor core. These pose some of the greatest risks to personnel. The overall aim of the exercise is to dispose of these wastes in a safe form such that they contribute zero, or virtually zero, risk to the environment over the very long time needed for their radioactive decay. This approximates to ten times the radioactive decay half-lives of the isotopes concerned.

The general principles against which the wastes are disposed are that (i) additional radiation doses to the population are minimized, (ii) damage to living species in the environment and restricting access to future mineral resources does not occur, (iii) permitted dose levels are not exceeded, (iv) multiple safety barriers are employed and these ought to be passive, *i.e.* not dependent upon servicing, (v) monitoring ought not to interfere with the passive multiple barriers, and (vi) liabilities ought not to be passed on to future generations.

Routine disposal of some low level wastes (LLW) commenced at the British Nuclear Fuels site at Drigg in Cumbria in 1959 and some wastes were disposed of into the deep ocean between 1949 and 1982. The Royal Commission on Environmental Pollution's 6th Report in 1976 led to the establishment of a Nuclear Waste Disposal Corporation (now NIREX) and a Nuclear Waste Advisory Committee (now RWMAC and with a contribution from ACSNI).

Thus, the job of disposing of intermediate level and some low level waste has been allocated to UK NIREX Ltd. It catalogues on an annual basis the amounts of waste in store and makes predictions concerning current and future arisings. Plans for the disposal of LLW, ILW and HLW (low, intermediate and high level wastes) are being made with the former two being the purview of NIREX. Figure 19.5 indicates the volumes of these different types of waste, but many will find that it is far more instructive to consider William Waldegrave's request that these volumes be depicted in terms of blocks and offices forming the Department of the Environment when he was then Minister (1985) (Figure 19.6). In terms of a readily understandable description of the activity of

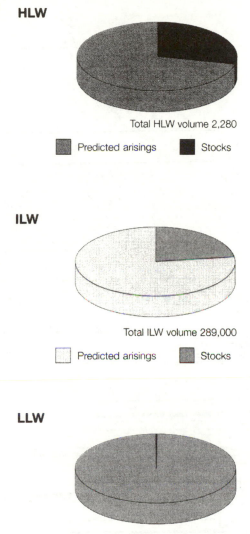

Figure 19.5   *Cumulated predicted arisings of radioactive waste in the UK after condition-
ing post year 2060. The current wastes in 1994 are also shown. Volumes are
in* m³
(*Source*: NIREX Report No. 698, June 1996)

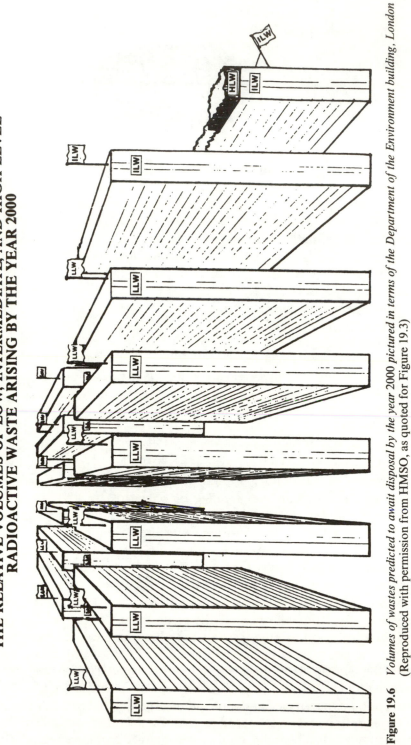

THE RELATIVE VOLUMES OF LOW, INTERMEDIATE, AND HIGH LEVEL RADIOACTIVE WASTE ARISING BY THE YEAR 2000

Figure 19.6 *Volumes of wastes predicted to await disposal by the year 2000 pictured in terms of the Department of the Environment building, London* (Reproduced with permission from HMSO, as quoted for Figure 19.3)

these wastes Lord Marshall, Chairman of a nuclear power generating company, the Central Electricity Generating Board (CEGB), introduced us to 'the garden unit' in 1986 (see Appendix).

Whatever the route of disposal, it must be realized that our environment is a precious asset and cannot be used up by our generation; rather, it must be sustained for the use of future generations. Also, the disposal site and waste packaging must be publicly and internationally acceptable. High level wastes will be stored for 50 or more years in vitrified form within stainless steel canisters awaiting future disposal in a location deep down and yet to be identified. Intermediate level wastes are destined for an underground repository that is currently being researched by NIREX. Low level wastes may be used as back-fill material and for packing in such a repository or, as at present, disposed of into licensed surface repositories. Some solution materials are diluted into the oceans from a pipeline, and gases (for example, radioactive carbon dioxide, methane or tritium oxide) are released into the atmosphere but these routes are now being discouraged by the authorities.

Wastes of somewhat lesser magnitude are released into the atmosphere as gases from fossil fuel fire-powered stations or are diluted and dispersed into rivers and oceans, but it must be stressed that this route is against the House of Lords Environment Committee's recommendations that 'dilute and disperse' routes should be phased out in the future. In terms of radiological threat, these environmental impacts are extremely small. In terms of perception, they are perceived as being a significant hazard and in order to keep apace of escalating public anxiety permitted exposure limits have been reduced approximately once per decade for half a century.

For every method of disposal the law requires that a safety case based upon risk assessment and a range of scenarios for the specific sites concerned is submitted to the Department of the Environment and to the Nuclear Installations Inspectorate. After being appraised by independent experts, licences to research underground, to dig a repository and to commence actual disposal of radioactive waste may eventually be obtained.

Much of the safety case based upon dose and risk depends upon the chemical speciation throughout the route from waste, through packaging involved, and through geosphere to biosphere to life forms. At first sight, it appears that the maximum rate of flow for any waste leaking from the vault is determined by the rate of water flow in that area. In practice, the real speed is proportional to the chemical speciation and absorption. On the one hand, precipitation can take the radionuclide out of the equation and set the speed to zero. On the other, organic molecules (collectively called ligands), such as are obtained from the degradation of plastics or from the EDTA or NTA used as decontaminating agents in the nuclear industry, can considerably increase the solubility of a heavy metal radionuclide and, therefore, increase the net speed at which it travels through the geosphere to the biosphere. A particular challenge in the UK and other countries such as France and Japan is the fact that the spent nuclear fuel is reprocessed for recycling. This came about because of the three-fold increase in the cost of uranium ore between 1972 and 1977 (Figure 19.3). The advent of recycling brought the world market costs of uranium more or less back to baseline levels. The reprocessing of used fuel rods produces a significant amount of heat and highly active waste. Other countries which are disposing of spent fuel rods without recycling are not faced with the costs of reprocessing wastes, although it must be stressed that the used fuel itself is some of the most dangerous radioactive material known to scientists.

The entrapment of this high activity level waste (HLW), composed of fission products and the last vestiges of uranium and plutonium, in borosilicate vitrified form is a safe means of ensuring that it is not water soluble and will not react with materials in the vault or geosphere. Were a *whole* lifetime's supply of energy (some 3 kWh per 70 years lifetime) raised by nuclear reactors (in fact, only a *fraction* is raised this way), then the total amount of vitrified waste per person life to be disposed of would be approximately the size of a shoe polish tin or cigarette packet.

Low level (LLW) and some intermediate level waste of short half-life (SLILW) can be built into concrete and sealed in 200 litre steel drums. This cementation being alkaline has the advantage of maintaining a high pH (approximately 10–11) which reduces the solubility of metal ions as oxides or hydroxides, and such low solubility protects the waste from any groundwater that leaks into the vault. There are as many approaches to storing and disposing of these drums of concreted waste underground or near the surface as there are repositories in the World.

There is the interesting perception challenge as to whether we regard waste as a future resource that is to be stored retrievably (just as the spoil tips from gold, silver and other precious metal mines are now profitably being reprocessed to extract the last few dregs), or whether we lock the waste away in a non-retrievable form (*e.g.* within concrete tombs) so that bounty hunters, intrusive trouble-makers and even prospectors of a future generation cannot readily break through and release the radioactive material.

It must be stressed that each and every nuclear waste processing facility and every recycling facility will produce wastes in each of the three categories of LLW, ILW and HLW, and all have distinctly different compositions. Not only will the range of fission products and other wastes vary, but also their amounts present. Nevertheless, modern science can rapidly compile an inventory of these wastes and choose a packaging and vault material which chemically and physically locks them all away. These ILW and HLW wastes usually comprise up to 50 fission products and up to ten of the actinides, amongst which will be traces of plutonium. The Press regards plutonium as a headline element although, in practice, the reprocessing technology tends to maximize the removal of plutonium because it may be usable in other reactors.

The threat of chemical toxicity and of carcinogenesis from plutonium has been compared with other threats by the three routes of intake; the skin, inhalation into the lungs, and ingestion and absorption through the intestines (Table 19.1). This indicates that at least three materials are considerably more lethal and produce death more instantaneously than plutonium. It is astonishing to note that up to 1 g of plutonium may be swallowed with still only a 50/50 chance of carcinogenesis some 10–20 years later being caused by this source of radiation within the body.

Since the nuclear bomb holocausts of 1945, the World population has had a tremendous fear of matters nuclear, be they nuclear power or even the waste arisings. There is a fear of cancer, and of genetic changes, and perception tends to be based as much, if not more, on emotional rather than on rational argument. Sir Denys Wilkinson stated, 'The perception of a hazard, even if it has no basis in reality, can cause real fear, and anxiety, in the minds of local inhabitants and can blight the quality of life as surely as a real environmental hazard.'

Although state-of-the-art research and chemical speciation and transfer modelling approaches have been applied to assessing the risks of radionuclides eventually leaving the vault and reaching the biosphere via the geosphere and then on to humans, the real significance of these hazards has yet to be convincingly communicated to the public to allay their anxieties.

The chemical speciation of all naturally-occurring elements and ligands, and of the radionuclides and other associated wastes at all stages of this process, has to be carefully assessed and built into extremely sophisticated computer models. Figure 19.7 shows a very simplified form of plutonium in contact with carbon dioxide and water. Of the eight solids present and shown as boxes, all are insoluble, but it will be noticed that the plutonium ion when carrying a $+4$ charge has a myriad of solution complexes, all of which are positively charged. This speciation knowledge, which comes from well-checked models, would have been useful at the time of building the Windscale pipeline in the 1950s; oceanographers are well aware that the beds of the oceans tend to be negatively charged and so any slight traces of plutonium inadvertently entering the pipeline would not be widely diluted and dispersed by the Irish and Atlantic seas, but rather would stick to the micas on the ocean

Table 19.1   *Toxicity of plutonium compared with other toxins. The three routes into humans are ingestion, through breaks in the skin, and inhalation. The plutonium is a mixture of plutonium isotopes as oxide which is five times more toxic than plutonium-239*

Toxin or poison	Lethal dose (mg)	Time to death
Ingested (swallowed)		
Anthrax spores	<0.0001	–
Botulism	<0.001	–
Lead arsenate	100	hours to days
Potassium cyanide	700	hours to days
Reactor plutonium	1150	over 15 years
Caffeine	14000	days
Injected		
North American coral snake venom	0.005	hours to days
Indian king cobra	0.2	hours to days
Reactor plutonium	0.078	over 15 years
Diamondback rattler	0.14	hours to days
Inhaled		
Reactor plutonium[b]	0.26	over 15 years
Reactor plutonium[b]	0.7	3 years
Nerve gas (Sarin)	1.0	few hours
Reactor plutonium[b]	1.9	1 year
Reactor plutonium[b]	12	60 days
Benzpyrene (1 packet of cigarettes per day for 30 years)	16.0	over 30 years
	Lethal concentration[a] (mg m^{-3})	
Inhalation Atmosphere		
Reactor plutonium[b]	0.026	over 15 years
Reactor plutonium[b]	1.3	60 days
Cadmium fumes	10	few hours
Mercury vapour	30	few hours
Phosgene	65	few hours

[a]Four-hour exposure; [b]Exposure at different levels
Source: Nuclear Power and the Environment, American Nuclear Society.

floor. These could well become churned up by storms, tides and shipping, and eventually arrive on beaches which would dry out, from whence dust could blow inland. Thus, some 40 years after the pipeline was built we can now understand this classical sea-to-land transfer of material, and so it is not surprising that occasionally a few atoms of plutonium are found in the vacuum cleaners of houses near to the coast of the Sellafield plant.

The vast majority of conclusions and risk assessments directed towards predicting the future in terms of migration, threat and chemical speciation are now based upon internationally agreed models. These models must also have future predictive abilities based upon past history

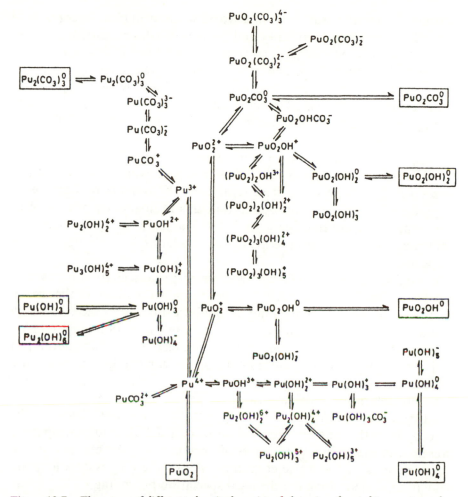

Figure 19.7 *The range of different chemical species of plutonium formed in a system when just plutonium, carbon dioxide and water are present. Solids are shown as boxes and solution species as formulae*
(*Source*: J.R. Duffield and D.R. Williams, *Chem. Soc. Rev.*, 1986, **15**, 291–307)

and current facts. The public perception of predictive modelling is becoming more acceptable as very many areas of design, planning and financial predictions now rely upon suchlike. It is far better to make the mistakes on the drawing board (in real terms within the computer) rather than physically to construct, try out and learn from the mistakes.

Regardless of whether we completely switch off nuclear energy and then honour our nuclear generation's responsibility of decommissioning all the power stations and disposing of the backlog of waste, or whether we remain as nuclear energy users or even expand its proportional contribution in the future, there will still be radioactive wastes for safe disposal. Even in a world without nuclear energy, there are still the research laboratory and hospital wastes requiring disposal. Thus, it is important for us to consider the World as it is rather than as it might have been if we had not turned to nuclear energy after the discovery of fission in 1939. Our responsibility is to leave this planet in a better state than when we inherited it, and this implies the safe storage, packaging and disposal of all radioactive waste as well as the chemical waste.

In spite of knowing the amount of waste awaiting disposal on the NIREX Inventory, there is still difficulty in perceiving how much waste arises from each person's lifestyle in the UK and his or her energy requirements, and the contribution that this radioactivity will make to our general exposure. The late Lord Marshall and his colleagues addressed this challenge in terms of Garden Units – see Section 21 and Appendix.

It is critical to realize that, whatever the disposal scenario proposed, it must be acceptable to the public in general, and to residents of the area in particular, as well as having international acceptability. The Chernobyl plume has proved that, in the event of a leakage, radiation does not stop at international boundaries and wait for visa clearance! Many developed and developing countries are now using nuclear energy and the more advanced countries that are already disposing of nuclear waste into the environment, or keeping it in a safe store, have much to teach those who are just entering into the nuclear energy era. Such lessons will address the perception of the different chemical forms in which the radioisotopes are present and the complex influences of groundwater, the geology and of the packaging of the waste concerned. All of these considerations are heavily dependent upon the chemical speciation prevailing.

Chemical speciation is defined as the chemical form or compound in which an element occurs in a living system or the environment, and can refer to solids, liquids and gases, in addition to complexes and interactions with naturally-occurring materials such as humic and fulvic acids. Figure 19.8 indicates more details of this concept. To quote a simple example, inorganic mercury as a dental amalgam is relatively harmless, whereas organic mercury compounds are highly toxic. On the other hand, organic arsenic compounds are less of a threat than the inorganic arsenic compounds which have been used to cause sinister deaths for centuries.

The determination of the chemical speciation at exceedingly low concentrations is one of the greatest challenges to analytical chemists of modern times. Techniques such as phase separation coupled with anodic stripping voltammetry and ion plasma analysis can analyse down to levels of parts per billion or lower. When metal ions are involved,

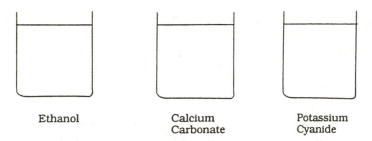

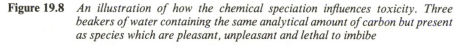

Ethanol   Calcium   Potassium
                 Carbonate   Cyanide

Figure 19.8   *An illustration of how the chemical speciation influences toxicity. Three beakers of water containing the same analytical amount of carbon but present as species which are pleasant, unpleasant and lethal to imbibe*

chemists will be aware that it is possible to establish a readily exchangeable equilibrium and so any efforts to try to concentrate the sample in order to make it analysable may well completely upset the equilibrium and produce the results for a non-representative set of circumstances. Thus, techniques which do not disturb the equilibrium are preferred. Computer simulation has many advantages in this respect as long as the software, databases and models are well verified and validated.

By and large, the approaches to disposing of all radioactive waste are similar in that they attempt to seal the radiation and toxic materials away from humans, albeit of a future generation, at least until the radiation has decayed away. Both used fuel rods and their highly active heat-generating wastes are somewhat difficult to handle and package because of the high temperatures generated and so plans aim to store them at surface level until cool enough to package and dispose of at a date half a century or more into the future. Naturally, surface storage and the ensuing radwaste vitrification processes are carefully licensed and monitored to minimize doses to operators at all stages.

Intermediate level waste, having been packaged, may be disposed of as soon as a repository is available. In the 1990s research focused on a site below the Sellafield region (in Cumbria) and a great deal of borehole research was undertaken to establish possible migratory pathways for leakages through the packaging and out from the vault in many hundreds of years time, which could eventually migrate towards the surface used by future generations on this planet.

Low level waste often has activities close to background levels but it is still regarded as suspect since it has come from a nuclear establishment. Tissues, gloves and other objects of this activity could, in principle, be disposed of into surface tips such as existed at Chatham dockyard and in current use at Drigg (near to Sellafield) and Dounreay (near to Thurso). Some of the future arisings of low level waste may well finish their days in a deep repository as spacing or as back-fill.

Subsequent to the borehole research, NIREX was hoping to build an underground laboratory (a Rock Characterization Facility, RCF) some 700 m below Sellafield with a view to following the research with an underground repository for ILW and some LLW. It was important that all data leading to such disposal were expertly assessed in terms of the science and its perception.

The latest independent appraisals were by a group of Royal Society scientists in 1994, by the Planning Inspector and by the RWMAC report *Rethinking Disposal 1998*.

The 1994 Royal Society Report on NIREX's proposal for the disposal of radioactive wastes in deep repositories mentions the international moratorium, which is due to be reconsidered in 25 years, and which currently precludes the dumping of radioactive waste into the seas. There was a perceived need to optimize interactions between scientists and design engineers and also to commission more modelling of combined gas, water and heat flow characteristics which, as of that time, had not been attempted. The Royal Society had previously made an important point in a report in 1992 indicating that ignorance tends to widen the gap between perception of risk and scientifically calculable levels of risk. Whenever the public has reason to suspect that there is a conspiracy to suppress public knowledge, their own anxieties may well be enhanced.

Since the Royal Society drew attention to the fact that NIREX had much of their documentation labelled 'commercial in confidence', NIREX are now publishing new details of the exact databases, parameter values and data which were previously shown only as diagrams. They also set aside time for thorough peer review of all the science.

The 1994 report also suggested that the future NIREX Research and Development Programme should consist of two parts. The first would be, as at that time, aimed at the study of short-lived intermediate level and low level wastes, and the second would investigate a related repository for long-lived ILW and, possibly, HLW as projected into a future generation. At that time, essential research was not going on in the UK into the disposal of HLW.

From an earlier visit to Sweden the Royal Society group had found a clear sense of strategy for the disposal of *all* types of waste. Recommendations to NIREX were made for further attention concerning the timetable, the two strategies just described, peer review, long-term evolutions of future repository sites, groundwater movement studies, and international participation. The Royal Society maintained an open mind as to whether Sellafield was indeed the correct site for a repository and offered the above recommendations in order to provide the best opportunity for a successful outcome of the scientific work being undertaken by NIREX.

Whereas scientists applauded this frank, constructive, independent appraisal of the NIREX research, the public was discouraged to hear that learned scientists had ostensibly disagreed. Clearly, all scientists must strive harder to put across the message that only through this kind of discussion can the best way forward be achieved.

Another critic of NIREX's less-than-perfect trust-building activities was Sir John Knill, a previous Chairman of the RWMAC who implied that the nuclear industry apparently concealed from Government nuclear waste disposal experts for five years how it came to choose Sellafield as the preferred site for burying radioactive waste. He stated that, 'There was a great lack of clarity concerning why Sellafield was not in the original five hundred options used as the original NIREX survey of sites. Rather, it was added in 1989 because it would be cheaper to operate but this was not made known at that time.'

Against this background NIREX applied for planning permission to construct a Rock Characterization Facility (RCF) at Gosforth near to the Sellafield site in Cumbria. This Rocklab would be similar to those used in other countries (*e.g.* Stripa in Sweden, Grimsel in Switzerland and Tono in Japan, *etc.*) to quantify rock stratification, pressures and water flow characteristics. Powerful arguments were put forward from both sides in terms of both the science and the perception.

The 1996 Planning Inquiry's Inspector's Report was submitted to the Department of the Environment and the then Minister, Mr John Gummer, announcing on 17 March 1997 that permission was being refused on the grounds of it being 'seriously premature', based on 'inadequate knowledge' and giving cause for concern about 'scientific uncertainties and technical difficulties in the proposals'. Other reasons for refusal included poor design and layout, and access arrangements on the surface being in a national park.

Cynics commented that the approaching May 1st General Election (which was announced the same day) swayed the decision, and quoted the Government's withdrawal in 1987 from four potential sites in Eastern Britain, also on the day a previous election was announced (they were all in Conservative seats!).

Assiduous commentators such as Professor Blowers of Social Sciences (Planning) at the Open University suggested that "British Policy is to ensure that radioactive wastes are not created unnecessarily and are safely disposed of at appropriate times and in appropriate ways. With no disposal solution in sight, the only sensible course is to stop more waste accumulating. Logically this means stopping all reprocessing – which is the biggest waste creator – and ultimately the entire industry. Looked at this way, loss of the NIREX repository may turn out to be the point of no return for Britain's nuclear industry."

The bad news was that more than £400 million worth of research was written off and that Britain fell further behind other nuclear countries in not having the next generation of radioactive waste disposal sites under construction. The good news to the waste generating companies was that a considerably larger financial outlay to build the repository after using the proposed rocklab data had been put off for a couple of decades.

NIREX are now restructuring their plans for selecting a future site for a rocklab and maybe a repository having learned from past experiences.

A multi-arithmetic decision analysis approach is being used which places greater emphasis upon transparency. In the future, NIREX plan to get correct not only the science but also the whole process of its perception, peer review by independents, and even site-volunteerism (as used effectively in France and the USA).

The criteria for site selection will include processes recommended by the International Atomic Energy Agency and by the joint Radioactive Waste Management Advisory Committee/Advisory Committee on the Safety of Nuclear Installations (RWMAC/ACSNI). The Regulators will be involved at an early stage and criticism is to be invited from all concerned (both constructive and, if merited, destructive). It will take about ten years to resolve these issues and, interestingly, there is no guarantee that the Sellafield site will not emerge from the site selection process as a good option. It has been suggested that an Independent Commission be set up to oversee this new wave of initiatives.

In January 1998, the RWMAC published *Rethinking Disposal* after due consideration of events leading up to the refusal of permission for the Rocklab. They stated that there are strong ethical and scientific arguments which favour disposal of ILW rather than its indeterminate supervised storage. This does not preclude the use of surface storage prior to disposal. Government were urged to reaffirm a policy of deep disposal of ILW, within policies of sustainable development and self-sufficiency in the disposal of these wastes.

Once again, the site identification process was seen as a pivotal aspect which needed to be transparent, and acceptable to the public. Licensing Departments were urged to test public opinion about a single repository, as distinct to separate ones, for ILW and for HLW waste. For the first time in an official report, the thought was raised about the best practical environmental option being a central repository (presumably in Europe) serving the needs of a number of countries.

20   Modelling Doses and Risks

It is better to identify the mistakes on the drawing board (in practice this is in the computer modelling) than to build constructions in reality and

Figure 20.1 *Scientists inspecting an extinct Natural Nuclear Reactor at Oklo which is believed to be the oldest known natural reactor and, although extinct for millions of years, can be used to validate the migration of radionuclides as computed by sophisticated modelling*
(Reproduced with permission from *Managing Radioactive Waste in the European Community*, CEC, Brussels, 1994)

learn from bitter experience. These models are only as good as the data, and two-thirds of the research time spent modelling the future is dedicated to checking data and validating the model by, for example, use in analysing fission product migration from extinct reactors in nature or from laboratory research (Figure 20.1). The mathematics of the model is one of both probabilistic as well as deterministic parameters. The former refer to the likelihood of an ice-age, an earthquake, of concrete fracturing, *etc.*, whereas the latter refer to data that can be determined according to the laws of science, such as the solubility product of an oxide or the formation constant of a ligand with a radionuclide ion.

The previous paragraph uses the term *computer* modelling. It must be stressed that many of the problems inherited in the forms of decommissioning nuclear energy plants, warheads and chemical and disposal threats arose from pre-computer days when such scenario modelling was clearly not feasible. Indeed, most modern wristwatches contain more computing power than existed in the whole World in 1961 and our nuclear era commenced almost two decades before then!

Nowadays such computer-based models are widely used in high-technology research, be it in studying the beat of the human heart, the role of a tactical jet fighter, refining processes in the chemical industry or waste disposal systems. In addition to their predictive nature stretching forward thousands of years they have a distinct advantage in assisting the choice of the most cost-effective means of achieving today's research objectives. They raise the most important questions which need to be tackled at laboratory level in order to assist with further validation of the model and achieve a higher degree of reliability upon the figures generated by such computer programs.

Some years ago, the European Union co-sponsored a programme amongst many of its partners whereby 288 such chemical and geological modelling programs were compared – a final list of 15 acceptable models was agreed internationally for use in radioactive waste speciation and migration modelling. Similarly, a database was established covering most of the species formed from materials likely to be present as fission products or actinides. Later developments included building in devices for estimating the retardation arising from interactions with pores and the surfaces of rocks, and also the role of colloids in terms of keeping normally insoluble metal oxides mobile with the aqueous solution flow. Figure 20.2 compares a very simple model studied using different modelling programs, and the degree of compatibility between all of their output indicates why these programs were recommended by the EU assessors.

Many of the data required for such modelling are site-specific; for example, the geology, hydrogeology, weather pattern, influence of earthquakes and ice-ages, *etc.*, as well as the type of topsoil and source of drinking water for the population. This usually implies that applications to develop a repository must proceed iteratively.

There are many difficulties in setting up thermodynamic databases and these are surpassed only by the difficulty in setting up kinetic databases

Table 20.1 *Chemical analogues of plutonium, in its various oxidation states, which may be used to study its chemistry and biochemistry*

Element (oxidation state)	Analogue of
Lanthanide(III)	Plutonium(III)
Thorium(IV)	Plutonium(IV)
Neptunium(V) as NpO_2^{2+}	Plutonium(V) as PuO_2^+
Uranium(VI) as UO_2^{2+}	Plutonium(VI) as PuO_2^{2+}

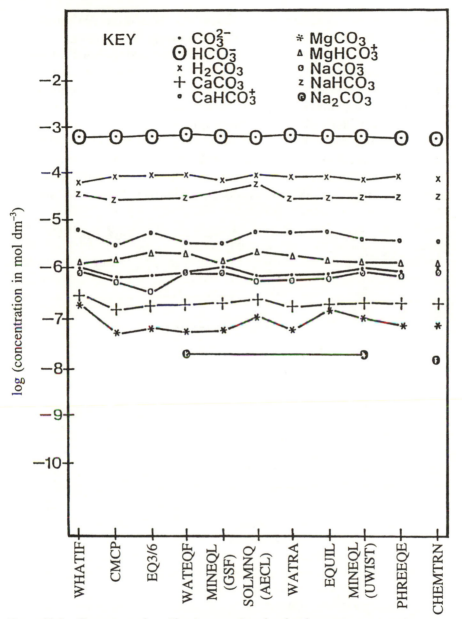

Figure 20.2 *Computer code verification exercise whereby eleven computer programs were used to calculate carbonate species' concentrations present at equilibrium* (*Source*: European Union CHEMVAL Project, 1990)

for researching slow reactions. Table 20.1 illustrates the wide range of oxidation states that one can expect to find for an actinide element such as plutonium, and how in the absence of reliable data it is necessary to build up a database by extrapolation from chemical analogues. Most of this exacting, but not necessarily exciting, process has now been completed, and internationally accepted databases exist for modelling radionuclides in the vault and in the neighbouring environment.

Conceivable pathways through which leaching from a vault will move towards humans of a future generation can be schematically depicted in different ways depending on whether one is a biologist, zoologist or physicist. The one shown in Figure 20.3 is as seen by a physical chemist who would consider a very small but finite volume somewhere between vault and human and then repeat this process many times over until the complete pathway had been considered. Within this volume, in addition to the radioactive decay of the isotopes concerned, a wide range of chemical and physical reactions can occur as shown. Thanks to the laws

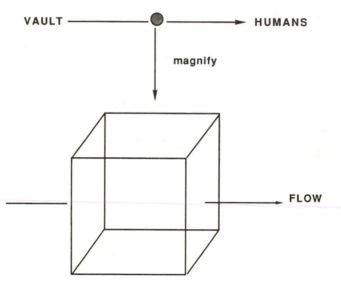

Processes considered for each "cell" include:

        Diffusion into neighbouring cells

        Precipitation/dissolution

        Isotopic exchange

        Physical adsorption

        Electrostatic adsorption

        Chelation

        Diffusion into pores

        Diffusion into solids

        Ultrafiltration

        Radioactive decay

        Chemical speciation

        etc.

Figure 20.3   *Illustration of how the physical chemistry of radionuclide migration and decay is modelled using computers*

of thermodynamics and physical science, it is now possible to speciate all of these processes chemically using the software and databases mentioned above and to calculate the species present at each point. These objective data can then be built in to the overall models using probabilistic considerations of how the material moves from the vault to the geosphere and then to the biosphere in response to natural events.

It must be stressed that any areas with which the modellers are not completely satisfied must be thoroughly researched. This is illustrated by Figure 20.4 which shows the range of metal ion complexes found in natural waters. The greyest area of certainty lies amongst the fulvic and humic acids and extensive research is now in hand to establish the criteria that determine their reactions. Another area that required deeper investigation was that of microbial action upon waste. Since microbes tend to live upon organic material and excrete potentially metal-mobilizing ligands, there is now a move to reduce the amount of organic material present in radwastes. Furthermore, when a cleansing agent such as EDTA (ethylenediaminetetraacetic acid) or NTA (nitrilotriacetic acid) is used for cleaning a surface, it has been shown that these ligands can permit metal ions to travel up to a million times faster through the geosphere than is possible in the ligand's absence. Bearing in mind that some 10 000 tonnes of EDTA and NTA are dispersed into the environment in the UK each year, metal mobilization in the presence of such chemically synthesized ligands is a serious challenge. The situation is exacerbated by the fact that EDTA is non-biodegradable and, therefore, has already built up to undesirable levels in some rivers.

The geochemical subject of natural analogues for checking the validity of models and the modelling process has been widely researched. Scientists now study a range of sites around the world where actinides, such as uranium, and fission products have leaked out of a mine or from a naturally-occurring nuclear reactor which is long since extinct (see Figure 20.1). A correlation between the groundwater composition and the suspected timescale of events is a good confidence-building exercise of relevance to our ability to predict the future of a disposal site in the UK. Similarly, existing low level waste disposal sites, such as Drigg, can be spiked with a small amount of low level radiation and the migration of that material through the topsoil and clay can be studied over a distance of a metre or so.

Another validation exercise is part of a series done at Harwell whereby concrete spiked with plutonium oxide was exposed to water at different pH values and then the predicted dissolution of the plutonium from the concrete was compared with that measured experimentally (Figure 20.5).

When demonstrating that the long-term prediction process works, much of the research focuses on plutonium-239, which is an alpha-

	Soluble		Colloidal			Particulate
	1 nm		10 nm	100 nm		1000 nm
Free Metal Ions (aq)	Inorganic Ion pair Organic Chelates	Organic Complexes	Metal Species Bound to High Molecular Weight Organic Material	Metal Species Adsorbed on Colloids	Metals Incorporated into Organic Particles and remains of Living Organisms	Mineral Solids; Metals Adsorbed on Solids; Precipitates, Co-precipitates
$(X^{(n+)})_{aq}$	XSO_4^{n-2}	X-fulvic acid	X-humic acid	X-Fe(OH)$_3$ X-MnO$_2$	X-organic solids	X-clay

C O L L O I D S

Real? .... Pseudo?

Figure 20.4   *Range of chemical speciation of metal ions in natural waters that are considered by computer modellers and an indication of sizes*

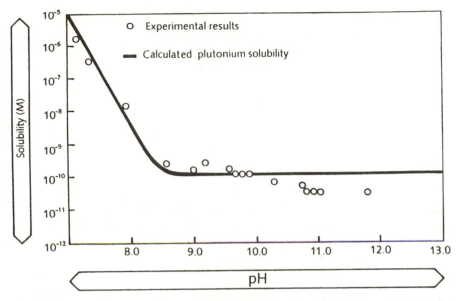

Figure 20.5 *Computer simulation of solubility of plutonium sealed in concrete exposed to water of different pH values. The points represent experimentally measured solubility and the lines are the calculated soluble species* (*Source*: UK NIREX Limited)

emitter with a half-life of 24 390 years. This suggests that it would take a quarter of a million years before its activity has decayed away to ignorable levels. (The rule-of-thumb is to use ten half-lives.) It also underlines the fact that even the most well-constructed disposal vault will long since have decayed away or succumbed to geological crushing pressures before the last few radioactive isotopes of plutonium disappear.

Thus, plutonium is studied in all of its oxidation states complexing with all of the ligands that may well be found in the waste, or in the degradation products from the waste, or in the geosphere and biosphere as it passes outwards. The prevailing conditions will be dictated by the packaging material, such as concrete, as well as by the redox scenario in and around the vault which is usually dictated by the iron-containing rocks therein. The oxidation and reduction of the plutonium will depend upon the air that gets through to the vault. Modern modelling techniques couple this speciation knowledge with adsorption and co-precipitation processes, with flow and mass transport phenomena, and with sensitivity studies to check for pivotal data which need to be redetermined.

In the biosphere, plutonium in the +4 oxidation state (the most common oxidation state) tends to have similar biochemistry to that of ferric iron. Thus, in humans, it tends to go for transferrin (one of the normal binding sites for iron) and it is also found complexed to citrate in the blood. Almost negligible plutonium +4 aqueous species exist at blood plasma pH 7.4 since precipitation prevails. The normal routes of a contaminant getting into humans apply, *i.e.* through cuts or skin-pinches in glove boxes into the hands of workers in nuclear establishments, from inhalation into the lungs, and from digestion as part of a dietary impurity. Much research has gone into designing ligand drugs that can remove plutonium from wounds and also the question of lavage of radionuclides from the lungs has been widely researched. In general, the best all-round agent for removing plutonium is DTPA (diethylenetri-aminepentaacetic acid), but its use is a skilled area of medical chemistry.

The removal of plutonium from humans is an exceedingly challenging area that requires collaboration between toxicologists, clinicians and radiobiologists. Other than the fact that some of these agents are alpha-emitters, it is often wise to leave heavy metals which have contaminated the body in the bone where they will do little harm. However, with radionuclides, the radiation being located so close to the stem cells of marrow is a major problem.

All of these parameters are brought together in a large variability analysis code which is used to predict the risk from each of these radionuclides and any toxic chemicals formed – this covers a wide range of species from cradle to grave, *i.e.* from the disposal package, through the vault, through the geosphere, through the biosphere, and then into humans and other species. The deterministic models arising from the chemistry are combined with probabilistic models. The modelling process is computed several hundred times per day and an overall average dose expected as a radionuclide threat is compiled from hundreds of such computations to produce an overall dose *versus* time plot. The one illustrated for the scenario shown in Figure 20.6 demon-strates that the Government's limit of keeping doses to within those giving a one in 1 000 000 increase in the individual developing a fatal cancer or a serious hereditary defect has been satisfied in this instance.

To conclude our discussion of this section, the reader will perhaps agree that, given high quality databases and sophisticated and inter-nationally accepted computer simulation software, it is possible to investigate scenarios which are too complex, too expensive or too experimentally-involved to be studied in the laboratory, to be able to predict far into the future and thus present a reasoned safety case. This is only half the challenge in that the next stage is one of communicating this safety case with inspectors, assessors, independent referees and, most importantly, with members of the public. Throughout it is important to talk to all interested and/or concerned persons and to point out that, whether nuclear power expands or contracts, there will still be much waste for disposal in terms of the backlog and current arisings, or in terms of decommissioning wastes in addition to that backlog.

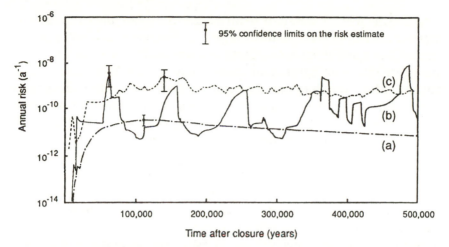

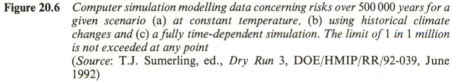

Figure 20.6 *Computer simulation modelling data concerning risks over* 500 000 *years for a given scenario* (a) *at constant temperature,* (b) *using historical climate changes and* (c) *a fully time-dependent simulation. The limit of* 1 *in* 1 *million is not exceeded at any point*
(*Source*: T.J. Sumerling, ed., *Dry Run* 3, DOE/HMIP/RR/92-039, June 1992)

21 Marshalling our Facts

The newcomer to the field would do well to build upon the outstanding lecture given by the late Lord Marshall in 1986, entitled *Nuclear Waste and NIMBY*.

The Marshall lecture and diagrams are reproduced as an appendix to this document; it was published in the nuclear magazine *Atom* of that year. To touch on the salient facts in précis form, Marshall points out that electricity is so widespread that we take it for granted but we also take for granted our abhorrence of the clutter arising therefrom in the form of pylons, power stations, *etc.*, being located in our backyards. The term 'not in my back yard' (NIMBY) applies to motorways, airports, industrial conurbations, *etc.* It is the natural reaction of any house dweller to be suspicious of new developments in their area. The wastes arising from nuclear energy are part of very many features of high technology which merit respect but certainly not fear. However, it is pointed out that in terms of perception the public fear nuclear waste because of its radiation risk. However, we, on this planet, have been subjected to radiation for 4500 million years. Marshall considers the damage that radiation does to human life and points out that every 1000 deaths encompass 250 attributed to cancer, of which one may be caused by radiation arising from background levels. Later, we shall see that the disposal of radioactive waste contributes thousands of times less than this one per 1000 arising from radiation.

Part of the fear of radiation is attributed to the fact that human senses cannot detect it, whereas instruments can detect it down to extremely low levels. This is an evolutionary feature since humans would surely have evolved methods of sensing the threat had it had been a meaningful challenge to us. We do marginally detect radiation from the Sun when we lie out of doors to acquire a suntan but otherwise radiation is invisible (except in the visible, audible and heat wavelengths). To a scientist, radiation is very easy to detect using a Geiger counter and such equipment has shown that the background levels giving rise to one in 1000 deaths vary from place to place. For example, it was mentioned earlier the radon gas emanating from the granite rocks of North East Scotland and of Cornwall. More recent reports list other areas where the rocks near the surface are porous and thus a larger volume of radon enters into the homes. Levels in some caves have been reported at several hundred times the action level for homes (200 Bq m^{-3}). So, too, on the Cornish beaches one can find pebbles far more radioactive – because of the uranium therein – than any contaminated piece of seaweed or other material ever found on the Sellafield beaches. Similarly, the luminous dials of old-fashioned watches, clocks and alarm clocks used radium as a source and this is far more radioactive than any of the beach materials. Nevertheless, we happily sleep beside these sources of radiation for eight or nine hours per night throughout the year. Gas mantles contain thorium but are widely used in the camping and leisure pursuit industries and, once again, they are located in the sleeping quarters of recreational vehicles and tents.

Marshall has calculated that the average exposure and risk from background radiation is equivalent to five puffs of a cigarette once per week. That is, one quarter of a whole cigarette (assuming about 25 puffs per cigarette) per week. This is down at the level of background secondary smoking inhalation similar to passing through a group of smokers huddled around the entrance to a no-smoking building or passing five times a week through a smoking-permitted carriage on a railway train. The dose is a radiation risk that we cannot avoid but we can vary the level of it by being in different parts of the country.

Marshall's radioactive garden refers to one-tenth of an acre as being the average sized garden purchased using a mortgage in the UK. The amounts of radioactive material are seen to be 0.8 kg of potassium-40, 6 kg of thorium and 2 kg of uranium as uranium ore. This material is distributed fairly evenly in the top 1 m of soil and has been there for thousands of millions of years. This radioactivity Marshall termed the 'shallow garden unit'. When discussing the disposal of intermediate level waste at a depth of more than 300 m (the depth contemplated in 1986), a similar 'deep garden unit' is defined. There, you have almost 3 000 000 kg of potassium-39 and -40, 2.5 tonnes of thorium and 0.8 tonnes of uranium.

In the days when Lord Marshall gave his lecture, the risks of passive smoking were not widely understood. However, since the phrase has now come into our everyday language, it is convenient to express the risk of death per annum from background radiation as being equivalent to that of passive smoking. Similarly, it helps to compare radioactive waste to a shallow garden unit for low and intermediate level wastes (short-lived) and to the deep garden unit for long-lived intermediate and high level wastes. The estimates used for the total amount of radioactive waste at these different levels have assumed that **all** a lifetime's electricity amounting to 300 000 units is nuclear in origin. In terms of volume, LLW is a suitcase size, short-lived ILW is a milk bottle size, long-lived ILW is that of a shoe box and HLW is approximately the size of a shoe polish tin or a packet of cigarettes. At this point it is wise to remember that only **one-fifth** of our lifetime's supply of electricity comes from nuclear power and so in reality any risk and volume as quoted above may be reduced by a factor of five for the average person's lifetime.

Bearing in mind that background radiation is responsible for approximately one in 1000 recorded deaths per year, we can now see the contribution of the various forms of waste towards this one in 1000. Low level waste contributes one-twenty-thousandth, short-lived ILW one-three-thousandth, long-lived ILW one-twentieth, albeit that this is of a future generation when the wastebearing groundwater comes from several hundred metres down to pose a threat (by which time much of the radiation will have decayed away) and HLW is initially four times the background level that one gets from radiation more than 700 m down. However, the fact that this waste is to be in vitrified form in stainless steel cylinders and buried deep down so that its radioactive decay characteristics are such that, after 100 years, amounts are below the level of the garden unit clearly reduces the relative dose. Even so, it is likely to be many thousands of years before the canister and some of its vitrified contents disintegrate and migrate back to the surface. It is clear that, by that time, it should not pose a threat that is unacceptable. In fact, much of this HLW will be stored for a century before being consigned to its ultimate disposal vault.

Finally, assuming that a lifetime's energy supply came from coal-fired power stations rather than being of nuclear origin, this energy is equivalent to about 20 m^3 of power station ash. Interestingly, the radiation contained in that coal ash is about a tenth of the garden unit and has a very long half-life and so would persist as a threat for longer than even the HLW during the first 100 years. Nevertheless, recognizing that coal-fired power station ash is still a low level of radiation, it is not disposed of at great depth like the radwaste that it is, but rather used in building roads and making breeze blocks to build walls for our homes! The fact that we have been obtaining power from coal for centuries and have become familiar with the waste therefrom, and that we are quite happy to use it in construction indicates the perception gap that needs to be closed between coal and nuclear power.

Since Lord Marshall's lecture there has, of course, been the Chernobyl incident and this can be used to illustrate how large populations can numerically lead to statistical deaths arising therefrom. Chernobyl released small particles of radio-caesium and radio-strontium which travelled long distances across Europe. Public understanding of the consequences of the Chernobyl incident is often confused because radio-activity is invisible and is little understood as a pollutant. Further, its side effects can be delayed before onset for many years. There were 29 workers who died from acute radiation sickness at Chernobyl and there were 14 subsequent deaths which may well be attributed to exposure from the incident. There were many other observable effects such as the development of treatable thyroid cancers in several hundred children living in the affected areas. Some 100 000 were permanently evacuated from the exclusion zone of 30 km around the Chernobyl plant. Some 25% of the gross domestic product of Belarus is being devoted to treating the Chernobyl incident.

Bearing in mind the risk of death is one in 1000 from background radiation and one in 20 000 arising from LLW, *etc.*, the reader will see that anything that raises the background level will statistically lead to an increase in deaths proportional to the size of the population considered. Some year and a half after the Chernobyl incident, visitors from the UK Electricity Generating Board visiting Chernobyl toured the site and examined the remains; using personal monitors the visitors were found to have had an increased dose arising from the tour of 30 microSieverts (μSv) = 0.03 milliSieverts (mSv). However, the total increased dose above background that they had received during the aeroplane flights to and from the Ukraine was 21 μSv. This is the usual dose we acquire for a flight to a holiday resort or a business trip elsewhere in Europe. The remarkable reduction in the level of radiation around Chernobyl achieved during those 18 months by decontamination measures and through natural processes has not been reflected in the difficulties that scientists have in demonstrating that much of the area around Chernobyl is now safe again. An indication of the scale of radiation doses quoted in milliSieverts is given in Table 21.1.

Indeed, at Chernobyl two elderly females refused to be evacuated and hid, only to be eventually discovered many weeks later and then moved outside the 30 km zone. As far as can be established, they are now

Table 21.1 *Typical radiation doses received by an individual in milliSieverts (mSv). The unit of potential biological injury is the Sievert which is the absorbed dose multiplied by the quality factor for the type of radiation to which the tissue is exposed*

Average total dose per year in UK from all sources	2.5
Natural radioactivity in the air	0.8
Cosmic rays per year	0.3–0.4
From radionuclides within one's body from food and drink	0.4
Medical/dental X-rays	0.1
Natural materials used in construction	0.3–0.7
Return flight of some hours in both directions	0.04
Consumer products*	0.01
Fallout from nuclear weapons testing	0.01
Dwelling within 30 km of a nuclear power station	0.001

Guidelines

Natural background in Kerala, India	10.0
Maximum annual dose limit for radiation workers	50.0
Dose to cause nausea but not death	1000
Dose to cause death within a few weeks	10 000

*TV sets, watches, smoke detectors, tobacco products.

perfectly healthy and suffering no side effects from the incident. This anecdotal type of evidence can be replaced by statistical deaths as shown in the map of Europe (Figure 18.1). This takes the millions of the population and multiplies them by the increased statistical death toll over 50 years arising from the increased radiation from the Chernobyl plume having passed twice over our country. In the UK, there are computed to be a statistical 37 deaths spread over 50 years arising from this increase in radiation which ought to be compared with 7 000 000 cancer deaths in total over the 50 years arising from background radiation. (There is an annual **fluctuation** of 6000 deaths per year!) Also, there is doubt about the ICRP figure of 10–12.5 deaths per mSv per 1 000 000 exposed which was the pivotal parameter used in the calculation. Clearly, the people who are the unlucky 37 will never be aware that the particular isotope that initiated their demise originated in central Ukraine.

Not all radioactive waste is disposed of below surface. A very small fraction is sometimes washed down effluent pipelines. Naturally, there have been allegations that this has raised the level of radioactivity in the Irish Sea to unacceptable figures. One cannot define an area of the sea that belongs to us like the garden unit and so we have to fall back on the idea of a kilogram of sea water (approximately one litre) (Figure 21.1). Sea water contains salts and these are radioactive. In fact, the average sea water radiation level throughout the world is 13.6 Bq kg^{-1}, some 88% of

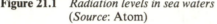

13.7 UNITS OF
RADIOACTIVITY
PER KILO

22 UNITS OF
RADIOACTIVITY
PER KILO

14.6 UNITS OF
RADIOACTIVITY
PER KILO

15 UNITS OF
RADIOACTIVITY
PER KILO

Figure 21.1 *Radiation levels in sea waters*
(*Source*: Atom)

which is from naturally-occurring salts; another 7% arises from nuclear weapons testing fall-out. The most radioactive sea is the Dead Sea because of the salt content which has 178 Bq kg^{-1}. However, turning to open seas, similar to the Irish Sea, the Persian Gulf has 22, the Red Sea 15, the Eastern Mediterranean 14.6 and the Irish Sea 13.7 Bq kg^{-1}. But what about the uranium that may leak from pipelines into the Irish Sea? In fact the total uranium discharge into the Irish Sea per year is 0.1 million million Bq. This compares with a natural uranium content of the sea in that area which is 200 million million Bq. In September 1997, the UK Minister for the Environment announced that a target sea emission of nil within 20 years was being adopted as Government policy.

The handling of the perception aspects of this waste disposal is now addressed.

22 Perception of Radwaste Disposal

It is amusing to contemplate a recently evolved hypothesis concerning matter:

> *All containers weigh more when empty and thus have*
> *to be discarded as litter as soon as possible.*

D.R. Williams, 1996

A corollary of this proposal is that it is someone else's job to clean up one's messes. Should you doubt these words, then look around your national parks, shopping malls and railway stations! Cynical as this thought may be, there is a serious side. If the general public are not interested in domestic waste disposal, the task of altering their perceptions in respect of radioactive waste disposal is orders of magnitude more difficult. Nevertheless, we must all start out on this long road to better communication and perception, and the following paragraphs hint at some ways of achieving these objectives without pummelling the uninterested public with too much information.

In the eyes of the general public the perception of radioactive waste is dependent upon the information that they are given and also upon their need to reach an opinion. By and large, most members of the public are not at all interested in forming opinions on such complex issues. If the issue is forced upon them because of an impending development in their neighbourhood, then they will usually vote for maintaining the *status quo* and not having the development there at all, unless there is an obvious advantage in terms of jobs and increasing prosperity for the area. They will not go out of their way to gain information from annual reports of learned Government experts (a recent annual report published by a Government advisory body sold less than ten copies!); rather, they tend to receive their information from newspaper headlines and from media articles which, in turn, attract a larger number of purchasers by adopting an anti-development stance and 'shock-horror-scare' headlines about nuclear material being smuggled across national borders and being available on the black market.

Some members of the public now receive substantial amounts of information from visitor centres such as the ones at Sellafield and other nuclear establishments. These are welcoming exhibitions and have hands-on facilities for youngsters with accompanying leaflets and booklets. The responsible visitor centres that the author has attended openly invite criticism from professional persons and consider removing any hint of prejudice or oversell if a *bona fide* complaint is made. Table 22.1 lists seven factors which influence the perception of the public concerning the risk of radiation. The first six of these factors can be tackled by increased efforts, financial commitment and sincerity in terms of communicating information throughout a child or youth's educational years

Table 22.1   *Factors influencing perception of radiation risk*

1	Fear of unknown
2	No understanding of a nuclear reactor
3	Threat of nuclear explosions (Bomb factory?)
4	Can it be nuclear weapons manufacture?
5	Dangers from disposal of radioactive waste and environment spoilage
6	Cannot renewable energy resources solve all problems?
7	Growth of professional objectors

Table 22.2   *Acronyms used in planning discussions involving radioactivity and/or chemicals*

• LULU	–	locally unwanted land uses
• NIMBY	–	not in my back yard
• NIABY	–	not in anyone's back yard
• NIMTO	–	not in my term of office
• BANANA	–	build absolutely nothing anywhere near anyone
• CATGROIN	–	cheapest available technique generating random or incomplete number
• CATNIP	–	cheapest available technique not involving prosecution
• BATNEEC	–	best available techniques not entailing excessive costs
• ALARP	–	as low as reasonably practicable
• ALARA	–	as low as reasonably achievable

as well as through visitor centres, *etc.* The seventh group is one that needs to be considered by applied psychologists, but most of us will realize that professional objectors working towards a particular goal, which is often political, will use the sharpest weapons available in their fight. The threat of cancer and the threat of genetic effects, both from radiation, are formidable weapons!

Over and above the perceived threat of radiation, there are the automatic knee-jerk reactions to hearing of any development in one's locality. Table 22.2 shows some of the acronyms used by residents and by politicians when considering their best line to opposing a particular development. Interestingly, all of these factors apply to non-nuclear developments. However, when a radioactive waste disposal site is considered, the fear of radiation is superimposed upon these worries.

To some extent, we are now living in an age where the packaging is more important than the contents. The view of the public, which is led by the press, is often more important than the detailed science of the proposal. As stated previously, this is partly the fault of scientists. The fact that they are vigorously discussing pros and cons of the science and technology of disposal leads to doubt in the mind of the public. We ought not to forget that we are talking about wastes. Our main aim is to dispose of them rather than make a whole new science out of the subject. Much of the scientific work that has been done over the last decade or two needs to be explained in the future to build public confidence.

It is interesting to note the pace at which nuclear matters now happen. Whereas nuclear fission was reported in 1939, the first commercial production in the UK occurred just 17 years later in 1956. However, because of communication and public awareness factors, it may well take far longer to introduce new science concepts at a level that the public can understand and accept, and this is especially true for nuclear science. Thus, more than 40 years on we still have little physical progress towards a Radwaste disposal facility. Communication leading to these objectives is equivalent to placing one's scientists in the public stocks. It is a notional lottery as to whether their views will sound sincere and be accepted by their audience. Setting up forums for all groups to mix and discuss these matters, interspersed with a certain amount of humour can help (as illustrated in Figure 22.1).

One of the most useful tools in liaising with the public is to have a fully trained workforce who are convinced of the safety and the worthwhile cause of what they are doing and who have at their disposal the means of communicating this to their neighbours, friends, family and members of the public that they meet. This means that diagrams and tables have to be available in user-friendly form to be drawn upon whenever the opportunity arises. Everybody working in the nuclear establishment must be perfectly clear in their mind exactly what a Becquerel is and, furthermore, what is a megaBecquerel. Useful examples include the facts that the Becquerel content of instant coffee is in excess of the level stated by law to be permissible for sheep meat, or that one cup of coffee is equivalent to about 60 Becquerels of radiation intake. Such quotes are all pieces of information that ought to be at the fingertips of nuclear workers. Another item of stock-in-trade that ought to be available is a list of the benefits from the use of radioactivity (Table 22.3). We cannot stress this list of benefits too often. We have to use it every time we have a dialogue with local authorities, and we have to stress it when we speak to inspectors as a more sympathetic inspectorate would be a distinct advantage. On the one hand, they insist on wastes being disposed of as soon as they are produced, but more communication with the inspectorate concerning acceptable routes would be welcomed. The fact that different inspectors expect records to be kept in a different manner to their predecessor is also disconcerting to the technical staff concerned.

Table 22.3   *Perceived benefits from nuclear power using radioactivity*

Nuclear power does not have reliance on large quantities of imported ores
Low air pollution emissions from nuclear power stations
Nuclear power is a low risk source of energy
Nuclear energy releases carbon-based fuels as a source of carbon for chemical industry
Nuclear industry has the healthiest workforce of all 340 000 employed in energy industry
Nuclear power generates only a small quantity of radwastes
Medical diagnostic techniques
Radiotherapy
Sterilization of operating theatre instruments
Fire detector sources
Industrial uses of isotopes
Geological uses of isotopes
Chemical and medical uses of isotopes

(*Source*: G. Fowell, *Chemistry and Industry*, December 1996, 956)

Figure 22.1   *Radiation and pollution in cartoons*

(*Source*: *ECOnews*, December 1989, p. 1)

(*Source*: Andrew Glendenning, *Chemistry in Britain*, 1997, 21)

Figure 22.1   *Continued*

"Who Are These Public"

(*Source*: Andrew Birch, *The Times Higher*, 31 May 1996)

Figure 22.1   *Continued*

Their guidance would be valuable concerning interpretation of the regulations as would any hints they could give concerning routes to reduce the costs of radioactive waste disposal. This matter is particularly important to small users who are finding that disposing of an isotope after use in a hospital or university laboratory is actually more expensive than the original purchase of the material concerned.

Another factor that needs to be stressed by all when dealing with the public (and with neighbours, *etc*.,) is that all 15 countries in the European Union have agreed that waste must not cross international boundaries. Any waste coming to Britain as part of spent fuel rods for reprocessing is returned afterwards, or its equivalent, plus the reprocessed fuel to the country of origin. This is an important point to be registered when having discussions with politicians who fear losing votes were Britain to become the nuclear waste dump of the world.

One interesting challenge to researchers into radwaste is to invite them to draft out the press release on the possible conclusions of their work two or three years hence. Unless they can envisage a crisp and clean method of putting these facts over to the general public at the beginning of their project, it is pointless them researching for the next few years in order to get the technology right. Obviously, the scientific detail of the conclusions will take some years to determine but a justification as to why we are doing this research ought to be clear right from the beginning.

No amount of video films, leaflets, exhibitions, *etc*., can exceed the communication on a one-to-one basis. It is also important to realize that the communicator has to tune in to the type of person with whom the discussions are occurring. Recent data on the perception of all sorts of risks, such as dust, ozone depletion, radon, *etc*., have shown that, by and large, females have 10% greater fear of an item being hazardous than males (Figure 11.1). We must put greater effort into defining what is being targeted with radioactive waste disposal. It sometimes helps in communicating with all members of the public to get away from the purely technical aspects and to try to defuse the idea that radioactive waste is uniquely hazardous and has unique risks to human life.

It is useful to compare the wastes from nuclear energy production with those from other means of generating electricity (as done previously). Similarly, the waste can be compared in terms of other non-energy-generating hazardous waste, both in terms of quantities, the lifetime of the toxic constituents, the toxicities, the legislative and regulatory requirements, and current and future disposal options and their environmental impacts.

The way that the statistics are reported determines the perception. If we were to state that every inhabitant in the United Kingdom were to die one-ten-millionth of a death per year (that is to say their lives would be shortened by a few seconds a year), then there would be no cause for serious complaint. However, if we state that five people a year will die, that throws a completely different complexion upon the subject because the listener understandably worries about being one of the five concerned.

It is likely that the risk of fatalities associated with generating electricity through windmills is significantly higher than those associated with nuclear energy. This is because the former is a low energy density source and, therefore, very large numbers of windmill groups need to be erected and the dangers of being injured or killed by a whirling windmill blade are significantly increased *pro rata*. The manufacturing of the windmills requires a great deal of material fabrication and so there are associated risks in the factories producing these products and, indeed, in their erection on-site. In 1980, the Health and Safety Executive quoted approximately 160 deaths per year per gigawatt of electrical energy from coal-fired power stations and, similarly, for oil-fired stations if one includes deaths arising from the sulfur and related emissions. For a similar amount of energy generated by a nuclear-powered station, 0.2 deaths per year are expected. Note that there is a difference of a factor of a thousand when comparing fossil and nuclear fuels. Similarly, setting aside the sulfur content, the risk per unit of electricity in terms of radiation to the public is several times larger from a coal-fired power station plume than a nuclear-fired power station. The risk is associated with what one is prepared to pay for safety. The more you pay, the less the detriment; but to make the detriment equal zero requires paying infinity. On the other hand, if you pay nothing, you have zero protection, which is completely unacceptable. Thus, you have to find a point at which the cost–detriment equation is minimized.

For males in their early sixties, the chances of dying within the next ten minutes are about one in 1 000 000. This is the same risk as drinking half a bottle of wine or taking an aeroplane flight to Paris. We do not worry too much about that level of risk. In the areas of radiation, a one in 1 000 000 risk corresponds roughly to the reference dose from the variation of natural background from place to place in the UK. The radiation from radioactive waste management in the UK is about 140 times less than this one in 1 000 000 risk. It is approximately the average radiation exposure we receive from luminous dials. Talking in terms of millions, during one hour a million million million million radioactive atoms decay away from the waste in store at Sellafield. We do not have to worry about these dead atoms any more!

Let us consider how eliminating a threat can pose another threat and, rather sadly, distract attention from genuine, far larger threats. Turning from radioactive waste, let us look at the way that the press and popular magazines now present public enemy number one around the world. It appears that people who subject others to secondary cigarette smoke are regarded as dangerous subversives. A scare recently arose concerning persons who chew on venetian blinds because of the lead content of such screens. When one looks at the statistics of motor cars, these suggest that they ought to be banned altogether since these killing machines are responsible for thousands of deaths per annum. And yet they symbolize the ultimate degree of freedom of society to travel wherever and whenever they wish.

In the United States, reducing the speed limit to 55 mph did not eliminate fatal road accidents. The implementation of airbags, another risk-reducing feature introduced some ten years ago, is now responsible for about ten child and seven adult deaths per year from amongst those hit by the airbags; when deployed, they travel at 200 mph even though the car may only be involved in a relatively low speed bump. Nevertheless, legislation has insisted that even more expenditure goes into these airbags so that mechanics will have to deactivate or reduce the forces therein and, as of 1999, all cars being built must carry a new, smart airbag which can recognize both the force of the collision and the presence of a child passenger.

The point is that before implementing new legislation or new ideas it is the job of communicators to announce these ideas and to defend them through open debate. Involving the public in environmental decisions is no longer a negotiable option. It is mandatory. Although consultation goes better when the public are talking from an informed point of view because they respond more adequately, it must be realized that this takes longer to reach an end-point. The nuclear industry is now leading the way in terms of informed debate about many matters and should be proud of this pole position.

The public will not willingly go out of their way to gain information in order to become involved in informed discussion. Rather, they must be paid to do their investigations. North American Indian tribes were paid many millions of dollars to consider having a radioactive waste emplacement and to undertake feasibility studies. Perhaps, and not unsurprisingly, after the due period it was found that they had accepted and spent the monies but conducted but few inquiries, all of which culminated in saying a firm 'no' to the idea of having a radwaste site anywhere near their encampments!

Developers now attempt to win over the public by suggesting indirect counter-measures (see Section 12) such as improvement schemes in terms of roads, benefits to health, schooling, leisure activities, *etc*. Thus, although the radioactive waste disposal is imposed upon them rather than being voluntary, they will get compensation in the form of the quality of life.

Smoky road vehicles and handguns account for about 300 000 deaths per annum in the World. Surely eliminating these is more important than number-crunching concerning Chernobyl.

Scandinavia, and in particular Sweden, has built up a long tradition of public information and of informed debate. One means of describing the danger of a material is to equate it to the volume of water with which it needs to be diluted before it can safely be used as table drinking water. This is known as the 'toxic potential' and has been suggested as a means of exchanging wastes arising from reprocessing of material from other countries with the equivalent amount of toxic waste being returned to the country of origin.

Much of the power concerning planning permission in an area lies in the hands of local councils. It is long since time that developers talked turkey to local councils, and it will be a time of celebration when a local council actively welcomes a development involving radioactive waste because they thoroughly understand all the ramifications and because they are convinced that they are going to benefit from this project.

The public must also be told that challenging results helps in the debate. Scientists are trained to challenge and then to reach a consensus of opinion. This is far more healthy than merely placing a 'fatwa' upon local nuclear industry.

Public relations used to be the linchpin in terms of public acceptability. It is now realized that public relations is only one fraction of the overall scenario of consultation and debate. Throughout these consultations, one must stress the benefits of nuclear power, of waste reprocessing, of waste disposal, *etc.*, wherever the opportunity arises. The public is worried about the bogey of privatization and whether insufficient monies will be laid down for the future in terms of disposing of HLW in 50 or 100 years time. This bogey is stirred up by the media who has an undoubted bias towards opposing all developments in order to create headlines and sell its products.

The nuclear and environmental inspectors must be expected to have firm standards and must be persuaded to implement them without bias and with a certain degree of common sense. Long gone are the days when the inspector was to be feared.

The previous UK Prime Minister, John Major will long be remembered for rebuking the press for being 'People like you asking that type of question!' This confrontational approach to debate is very much counter-productive and it will take many months and years to rekindle the co-operation and goodwill from that press. Unless one can move a development forward without the press being convinced, albeit not totally and across the board, one may as well not start. As was stated earlier, it is important to write the press release almost at the outset of a planned project. This contrasts with many government departments where learned bodies construct reports and then the press release is composed by the press office, who have little knowledge of the science or of the perception. Throughout, humour helps as part of relaxing in each other's presence and as part of the bonding.

Clearly, to residents the construction of a site to dispose of low level waste could well lead to intermediate level waste, then to high level waste, and could eventually go on to a new power plant being constructed in the area. Do the waste producing organizations address the flaws in this chain argument? Table 22.4 shows the types of criteria that would need to be satisfied before further new power plants are built in Britain. Clearly, the likelihoods of these occurring are very low indeed and are not in any way linked into the sites chosen for low level waste disposal.

A difficult topic to put over to the public is the fact that molecules are made of different atoms. Some of these atoms will be radioactive, which is the subject of this document. These atoms have the same radioactive decay characteristics regardless of how they are bonded or whether they are present in isolation, and whether they are present as gases or solids or as solutions. However, the molecules into which the atoms are bonded will have different chemical personalities and some will be soluble and will travel rapidly through groundwater whereas others will be insoluble. Some will be bioavailable, and will be taken up by living species, and some will not. Even small changes can markedly affect the chemistry of an element and we must remember that the chemical toxicity of a molecule must be considered with equal weight with the radiation threat of the atoms therein. For example, we swallow 200 g of barium sulfate for a barium meal X-ray contrasting agent, whereas 200–800 mg of barium nitrate or chloride is highly toxic, and possibly lethal. Just by swapping common or garden sulfate for equally well-known chloride or

Table 22.4 *Factors leading to the construction of new nuclear power stations in the UK*

A	Demand for electricity is greater than conservation + load management + renewables
B	Current nuclear plants to be release- and accident-free
C	Nuclear generated power is economically competitive (Carbon tax?)
D	New plants have a shorter construction time and faster pay back than previously

C + D requires Government support (votes?)

nitrate, the toxicity has changed by a factor of 1000 or so. Thus, we cannot glibly brand an element as friendly or dangerous, as toxic or non-toxic; we must look at the different chemical species that it forms in its lifecycle.

The need to dispose of radioactive waste can best be communicated to the lay public if one continuously reminds them of the manifold benefits of radiation used in research, radiotherapy, energy generation, *etc*. With a growing interest in transcendental meditation, parapsychosis and other invisible spirits, it ought to be easier to teach the science of this invisible radiation and, in particular, to indicate how easy it is to detect with modern instruments.

It is particularly important that all people working with radiation, be they in the nuclear industry, research or in hospitals, must be given sufficient information, not only to feel comfortable and safe in their own jobs but that they must also be positively encouraged to discuss radiation and such concepts as the 'clusters' of incidents of conditions with their friends, neighbours and relatives. Bearing in mind that there are in excess of 20 000 people involved in radiation or radioactive isotope-based industries in the UK, this is a sizeable communication force touching upon all branches of the general public. An ample supply of booklets and leaflets should be made available from independent bodies such as the National Radiological Protection Board, the Department of Health, the Department of the Environment, Transport and Regions, as well as those from the nuclear industries. These must use everyday comparisons, such as the doses from flying in an aeroplane, medical X-rays, to background radiation, *etc*.

It must be realized that imparting this information to members of the public will not be easy. Nevertheless, it must be made the implied responsibility of all persons working in the industry. Many of their attempts will be met with responses such as, 'Well, we would expect you, an employee of the industry, to say that anyway.' This ought not to dissuade the person from trying to communicate further and eventually many lay-persons will be won over to what is, after all, a true and eminently defensible position. The Sievert and the Becquerel can be translated into garden units of risks which, in turn, can be related to the likelihoods or probabilities of tall jugs spilling their contents, to riders toppling off bicycles, to toast landing buttered-side downwards on the carpet, *etc*.

There is a need for simple, important concepts such as risk and dose to be introduced at primary and secondary school level. Initially, non-radiation based concepts such as the risks of smoking, solvent sniffing, drug taking, medical therapy, *etc*., can be used. Later, the reality of merits and de-merits of different sources of energy and of the challenges of waste disposal in these different forms can be introduced. The mathematics of calculating statistical deaths arising from a plume of chemicals rather than radiation could be an important feature of GCSE syllabuses.

Everyone involved in science, be they teachers or company employees, would do well to remember that civilization has evolved significantly since C.P. Snow wrote about our two separate cultures. Science is now very much part of the world environment in which we live and so a detailed study of the environmental chemistry of nuclear waste disposal and of natural background radiation is necessary in order to clear up many of the waste disposal problems facing our society. Scientists involved in radiation not only have to solve these problems but must also diligently set about the task of explaining the problem and its solution to the non-scientist. Further, the body politic (to use the words of C.P. Snow) must interact with the body scientific, and the media is increasingly being used as a means of communication between the two. When radiation workers abdicate their professional responsibility to interpret, to translate and to explain their technology to these two cultures, there may well be a vacuum formed which can soon suck in hot air and wild rumours, resulting in selected data then being put forward by special interest groups.

By and large, the expertise and knowledge necessary for the safe disposal of radioactive waste are already available. The application and implementation of such disposal routes using that expertise are still being tackled. Communicating the need for disposal and of the associated risks, doses and safety aspects has hardly commenced.

All involved must be encouraged to go out and not only to do a truly professional job at waste disposal, but also to communicate the solutions and ways forward with the public.

23 Concluding Remarks

There are pleas from all sides for a greater emphasis upon studying the science of perceiving risk and upon putting more resources into understanding the concept of risk and 'what is safe?' Mr Ian Taylor, a recent UK Minister for Science and Technology, Sir Kenneth Calman, Chief Medical Officer at the Department of Health, and spokespersons for many industries frustrated by the UK not being able to compete strongly because of our being held back by the public's misunderstanding of the risks of a new development have all written urging greater effort.

Mr Taylor wrote in 1997 that, 'We live in a society today that is safer than at any time in history. Yet, as our quality of life and standard of living improves, paradoxically we are becoming increasingly intolerant of risks. Too often, confidence in the benefits to society from scientific developments can be undermined by a single incident that is of marginal scientific significance, but which seems to imply that science is threatening.'

Sir Kenneth Calman wrote, 'The most interesting but the most difficult aspect of understanding risk is the perception of risk by the individual. Risks, no matter how clearly defined can be interpreted differently by different individuals.' Kant said, 'We see things not as **they** are but as **we** are', and in doing so clearly describes the problem. It does emphasise the importance of ensuring that the public are full partners in the process of risk assessment and management, and it is only with such involvement that progress can be made.'

Counting humans as part of the environment, it is paradoxical that during the last decade there has been a tremendous uprising concerning our apparent ignorance of the needs of the environment. This is ironic because we have witnessed an unprecedented amount of environmental research data being made available during that decade. Perhaps the difficulty arises because there are so many experts giving voice to their opinions, and to the lay-person these opinions seem to be of disparate views. As pointed out by Szerszynski, Lash and Wynn, it is difficult for the public to tell the signal from the noise in all these messages. The distinguished sociologist, Ulrich Beck has commented that the old industrial society, whose principal aim was the production and distribution of 'goods' has now been overtaken by a risk society focusing upon the distribution of 'bads', *i.e.* blames. Another distinguished international commentator, Ernst Bloch has described 'technology as the enemy within'. The Brundtland report of 1987 attempted to swing the emphasis of modern technology away from being an environmental threat towards the long-term objective of sustainable development. This approach offers promise provided that industry, governments and lay-persons are willing to keep the topic alive and to make sure that it progresses in this desired direction. The 1987 report stresses that we can only achieve environmental protection at the expense of some human economic and social development. Sustainable development demands that the notions of global equity, justice and basic human rights are intrinsically a part of the environmental issue.

What then ought we to be doing about it? There are three facts to grasp:

(1) With the recent developments in widely differing areas such as human genetics, biotechnology, information technology, and food preservation and production, a rash of 'shock-horror-scare' stories in the press has had a disproportionate impact upon the public.

Scientists must learn that it is their role not only to make new discoveries clearly understandable to the general public, but that they must also get involved in the devices through which the public will evaluate whether their discovery is good news or bad.

(2) There needs to be a greater emphasis upon teaching and understanding of risk and of safety at all levels of education. Only through this will members of the public have the means of assessing the status of new developments and the related issue of risk in an informed manner. The hard facts are no longer enough. This is even true when they are indisputable and are accepted as established dogma. We must be fully aware of all the other issues that influence the public's perception of these facts.

(3) One of the most important words in this discussion is that of 'trust'. We must have trust in scientists, we must have trust in industrial and technology managers, we must have trust in politicians and government at both county and national levels. Clearly, these are extremely tall orders indeed!

Ian Taylor as Minister of Science suggested that we establish 'appropriate systems for measuring and assessing risk'. The aim of establishing such a scale would be that any actions taken are more proportional to the potential problems envisaged.

The UK Government has set up an inter-departmental liaison group on risk assessment, chaired by the Health and Safety Executive. It will be interesting to see how much industry and education are asked to contribute to the workings of this group.

One of the ways of demystifying the ways that scientists express risk and the probabilities of novel and unfamiliar events is to set up a scale for risk. The Minister suggests a Richter scale similar to that used in measuring earthquakes. Anyone who has lived in the earthquake areas of California along the San Andreas Fault will realize that risk scale discussions are in daily use and that the public living in the area in homes built on that Fault are quite familiar with the numbers concerned.

It is suggested that the statistics of all kinds of naturally-occurring events such as being struck by lightning which is one in 10^7 per annum, and the risk of dying from cancer which is one in 350 per annum, are used as benchmarks on this scale. However, it would appear that even this is over-simplifying the situation since one must specify the age of the individual before quoting the definitive risk of dying of cancer, as clearly this risk increases in later years and the average over a whole lifetime is embodied in the figure of one in 350. As always, the question of unfamiliar risks having non-fatal consequences is being contrasted with familiar risks having severe or deadly consequences. On the one hand there is the 'dread factor', such as chemicals or radiation triggering off cancers and, on the other hand, there are situations where they are fairly laid back concerning the risk since they feel in total control. Smoking ten cigarettes per day on average will carry a risk of death of about one in 200 per year.

Why are governments becoming worried about the question? Why are industries becoming worried about such questions? Why are scientists worried about the image of their research? The answers lie in the fact that science, technology and industry can only have maximum effect upon wealth creation and improvements in the quality of life if they are supported by the public who are prepared to accept the risks as well as the benefits concerned. Each scientist may be regarded as producing science in the form of a flower or bloom. But these flowers can only best be arranged into the most attractive bunch provided that the scientist has put effort into communication as part of the research concerned.

So many multi-author textbooks have finished with the quotation, "I gather a posy of other persons' flowers but the thread which unites them is my own – Editor." The time has come perhaps for us to acknowledge that the communication or thread relating research is a vital element of all research work and that the researchers themselves and their line managers must take responsibility for convincing people of the merits of the work. Scientists skilled in measuring physical data may need training in interpersonal and communication skills through their academic teachers and human resource managers. Perhaps we ought to be producing fewer different flowers but placing more effort into the thread that relates all of these specialisms into a well-coordinated way forward for wealth creation and better quality of life.

APPENDIX

Your Radioactive Garden

Nuclear Waste in Perspective

The Lord Marshall of Goring, Kt, CBE, FRS
Chairman, Central Electricity Generating Board

This Appendix reproduces in entirety Lord Marshall's lecture, entitled *Your Radioactive Garden – Nuclear Waste in Perspective*, with gratitude for permission from Lady Marshall, Magnox Electric plc and Dr Myrrdin Davies (Oxford)

Electricity makes a vital contribution to our lives – in the home, in industry, in medicine and in a thousand other ways. We have got so used to it that we take it for granted.

Now in order to guarantee adequate, safe and economical supplies the Central Electricity Generating Board must have transmission lines and power stations, including some nuclear stations. However, Mr or Mrs Public says: 'I want all the electricity I need but *no* pylons in my view, *no* power stations near my home – and above all, *no* nuclear waste in my back yard; put it somewhere else.'

This NIMBY (Not In My Back Yard) syndrome applies to many other things as well – motorways and airports are examples – but its application to nuclear waste is very special because people *fear* nuclear waste. I think that fear is unjustified. People need to respect nuclear waste but not fear it.

In the public mind, nuclear waste means radiation risk. That is basically correct, but the nuclear industry considers that the radiation risk from nuclear waste is tiny.

Why do we believe that?

Radiation

Let us begin by considering what we know about radiation. The world has been subjected to radiation for more than four thousand million years, ever since the earth was formed.

Most of it (about 73%) comes from the natural radioactive materials that are all around us and indeed inside us, but about 12% comes from medical treatment (such as X-rays) and 14% from outer space in the form of cosmic rays (Figures A1–A3).

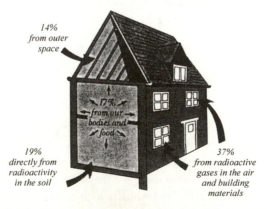

Figure A1   *Radiation is all around us*

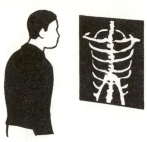

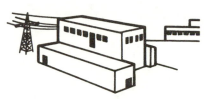

Figure A2   11.5% *from X-rays*

Figure A3   1.5% *from coal, travel, industrial use, etc., and the entire nuclear industry*

Before looking more closely at its origins, let me stress an important fact. We cannot sense radiation by smell, taste, touch, hearing or sight. I think that is basically why people fear it.

We can, however, detect it very easily with instruments. A Geiger counter, for example, is able to detect, display on a meter and give an audible signal to indicate the intensity of the background radiation around us.

This varies from place to place, according to the amounts of naturally-occurring radioactive elements which are present in the ground and in building materials.

The variation can sometimes be quite dramatic. For example, pebbles on certain Cornish beaches cause our detectors to indicate radiation levels many thousands of times higher than the normal background reading. That radioactivity is higher than you will find on any Cumbrian beach – and it doesn't come from British Nuclear Fuels' reprocessing plant at Sellafield; it occurs naturally. It is uranium ore.

High readings can also be obtained from old luminous dials on instruments, clocks or watches because radium (a radioactive derivative of naturally-occurring uranium) was used to produce the luminous paint.

Another naturally-occurring radioactive element in common use is thorium. This is used to make gas-light mantles and these, with the aid of a radiation detector, can readily be demonstrated to be radioactive.

So we are surrounded by radioactivity, and most of us are hardly conscious of the fact.

But why has mankind managed to evolve without the natural ability to detect radiation? Basically because it doesn't matter very much compared with all the other hazards that face us in our everyday lives.

It has never been a significant threat to our survival even in areas where the natural background radiation is well above average.

Five Puffs

In order to illustrate the risk to health from background radiation we can compare it with the risk from smoking cigarettes. There are indeed many similarities between the effects of radiation and the effects of smoking.

I can sum up all the work that has been done on this over many decades by saying that your average lifetime's radiation from natural sources – that is, from cosmic rays and from the radioactive material that is all around you – is equivalent, in terms of health hazard, to five puffs of a cigarette each week.

That degree of smoking is roughly equivalent to 'passive smoking' – that is, the involuntary smoking done by non-smokers in breathing in other people's tobacco smoke.

In the Soil

Now I should like to draw an analogy between the radioactive waste produced by the nuclear industry, and the natural radioactivity which is present in your garden.

In the top metre of an average-sized UK garden – occupying, say, one-tenth of an acre – is 7000 kg of potassium containing 0.8 kg of potassium-40, the radioactive form of potassium. There is also 6 kg of radioactive thorium and other associated radioactive elements. In addition there is about 2 kg of uranium, contained in what is essentially uranium ore. I mention the uranium ore because it also contains other radioactive products which make it more radioactive than the uranium we recover from the ore (Figure A4).

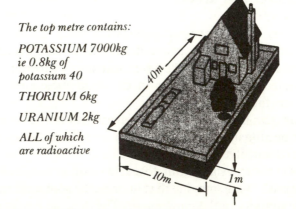

The top metre contains:

*POTASSIUM 7000kg
ie 0.8kg of
potassium 40*

THORIUM 6kg

URANIUM 2kg

*ALL of which
are radioactive*

40m

10m   1m

Figure A4   *A garden unit of radioactivity*

The top 300
metres contain:

POTASSIUM
2 800 0000kg

THORIUM
2400kg

URANIUM
800kg

300m

Figure A5   *A deep garden unit of radioactivity*

So all that very radioactive material is dispersed among the tonnes of earth that are in the top metre of your garden, and I shall call it a *garden unit* of radioactivity.

Let us now go down deeper and look at the natural radioactivity which is mixed up with the soil and rocks in the top 300 m. There you have 2 800 000 kg of potassium, 2400 kg of thorium and 800 kg of uranium. I shall call that a *deep garden unit* of radioactivity (Figure A5).

Real and Potential Risks

The radioactivity in the soil poses a *potential* and a *real* health risk. The top metre of the ground leads to a radiation dose, and hence a risk, due to direct radiation and from breathing radioactivity from it. The deeper layers do not actually contribute to direct radiation or to the dose received by breathing radioactive gas from the soil. The potential dose and risk from the deep garden unit is due to the possibility of the radioactivity getting into water supplies and food-chains and hence, eventually, into ourselves.

But what do I mean by *potential* dose and risk and how does it relate to *real* dose and risk? Well, imagine that in one year someone could breathe in *all* the radioactivity in the *garden unit*. The dose he would get is called the *potential* dose, but it is actually of the order of 10 million times

Figure A6   *The potential radiation dose*

greater than the *real* dose which is the annual dose from natural background radiation that he receives, because of the shielding and containment safety factors that nature provides. So the potential dose, and the potential risk, is 10 million times greater than the real dose and risk from the garden unit (Figure A6).

Now let us consider the deep garden unit. Of course the potential dose from the radioactivity in the top 300 m is very large indeed, but the real dose, and hence risk, is tiny – again because of nature's colossal safety factor.

So you have a large amount of naturally-occurring 'nuclear waste' in your garden. It presents a significant potential risk to you, though the real risk is a very slight one. That from potassium and thorium are each equal to about one puff of a cigarette each week and the uranium

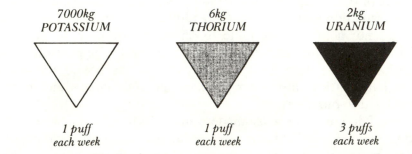

Figure A7   *The health risks from the radioactive materials in a garden unit related to puffs of a cigaratte*

gives a risk equivalent to that from about three puffs of a cigarette each week (Figure A7).

Let me now put an important question. Imagine that I have separated the radioactivity from the top metre in your garden and have concentrated it. Have I made it safer or more dangerous?

The answer is that it depends entirely on what I then do with it. If I eat it, or sit on it, I've made it more dangerous. But if I dig a deep hole in your garden and bury the radioactivity in concrete, I have made it safer, because I have removed it from your immediate environment.

Your Power: The Wastes

On average everyone in the United Kingdom uses directly or indirectly some 300 000 units of electricity during his or her lifetime. Let me suppose, just for the purpose of developing my theme, that all that electricity comes from nuclear power stations. (You may prefer all your electricity to come from burning coal. We will examine the consequences of that later on but I don't think you'll be too happy with them.)

Your lifetime's supply of electricity from nuclear power would give rise to various types of nuclear waste, some of which come directly from the reactor and some from the fuel reprocessing plant. It comprises some 16 l of low level radioactive wastes, half a litre of short-lived intermediate level radioactive waste, 4 l of long-lived intermediate level radioactive waste, and 0.14 l of highly-concentrated high level radioactive waste.

Let us consider each of these types (Figure A8).

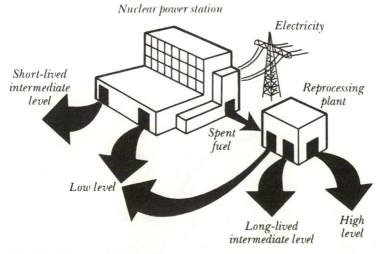

Figure A8 *The different types of nuclear waste*

Low Level Waste

First, low level waste. For an individual's lifetime nuclear electricity there would be 16 l of low level radioactive waste (about the size of a small suitcase), consisting of slightly radioactive day-to-day trash which arises in nuclear laboratories, power stations, and the reprocessing plant. It includes items such as paper towels, rubber gloves, glassware, air filters, floor coverings and incinerator ash. These items vary in radioactive content but their radioactivity is similar to many of the things around us which are naturally radioactive – commercial fertilizer and granite, for example – and all are much less radioactive than uranium ore and some luminous paints.

This is the bulk of the waste that the nuclear industry has and would like to dispose of. At present most of it is buried at a waste site owned by British Nuclear Fuels at Drigg in Cumbria, but in order to preserve that site for the low level waste from BNFL's Sellafield reprocessing plant (which contributes 10 of the 16 l I have mentioned) UK NIREX Ltd, the nuclear waste agency, is seeking an alternative site and facing fierce opposition – opposition which is saying 'Not In My Back Yard'. Of course, no one is actually proposing to bury waste in anybody's back garden. However, just putting emotions to one side for a while, why not bury the low level nuclear waste in your back garden?

Let us do a 'thought experiment'. Imagine that we take your total lifetime's nuclear electricity low level waste and present you with it, say

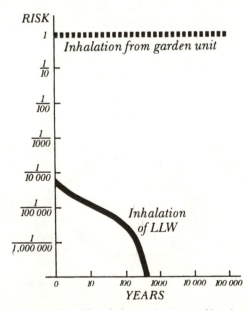

Figure A9   *Relative potential health risk from inhalation of low level waste*

on your retirement, and ask you to bury it evenly in the top metre of your garden. It isn't very much – 16 l: about the size of a small suitcase. So you spread it on the garden and mix it with the top metre of soil, which already, you will recall, contains a *garden unit* of radioactivity.

Obviously you have increased the potential radiation dose and risk associated with it – but by how much? This is an easy technical sum for us to do. The additional potential dose and risk is in fact 20 000 times less than was there already from the natural radioactivity (Figure A9).

That of course is only initially. The diagram above shows what happens over a period of time. The potential risk from the low level waste mixed with garden soil decreases dramatically because the level of radioactivity declines with time. That from the garden unit, on the other hand, stays constant since the radioactivity in uranium and thorium decays much more slowly than the radioactivity in the low level waste.

So if we do put your low level waste in your back yard the extra potential risk will be tiny. However, that was only a 'thought experiment'. We're not proposing to do that – we can actually do a lot better. What NIREX is proposing* is to use a very carefully chosen site (*not* in anybody's garden), to bury the waste some 10 m down (*not* mixed in the top metre), to place the waste in steel and concrete (*not* just tipped in the soil), and to ensure that the geology and hydrogeology are such as to provide extra barriers to prevent radioactivity from being released from the site. These measures will ensure that a potentially small risk is effectively eliminated.

Short-lived Intermediate Level Waste

But, you might ask, if low level waste is such a small matter, perhaps it is the other categories of waste that we should be concerned about. Well, let us look at the next category, the short-lived intermediate level waste.

For a lifetime's nuclear electricity, we each have a half of a litre of this waste (about the size of a milk bottle). It consists, for example, of metallic scrap, solidified sludges and resins, and it comes mainly from nuclear power stations. The radioactivity is significantly higher than that of low level waste: a Geiger counter would indicate levels similar to those from uranium ore. Wastes of this type are currently stored at power stations and other nuclear sites. They are, as their name implies, short-lived and they could safely be buried near the surface alongside the low level wastes. However, the plan is to bury them deep in stable geological

---

*Since the lecture was written the Government has decided that low level waste should be buried deep with intermediate level waste in a stable geological area. A typical depth would be about 300 metres.

areas well away from the human environment. A typical depth would be about 300 m.

As with the low level waste the public say 'Not In My Back Yard'. Well, let us do our 'thought experiment' again. Let us imagine that the short-lived intermediate level waste has been delivered to you and dug into the top metre of your garden. What effect has that on the potential dose and risk from your garden?

The answer is: very little. This is because the potential risk from the short-lived intermediate level waste is some 3000 times less than that from the garden unit of radioactivity already there (Figure A10). Moreover, the radioactivity gradually decays and even that potential risk disappears over the course of a few hundred years (Figure A11).

So I can again ask the question: 'Why not in my back yard?' or 'Why not in your back yard?' In fact it would be safer and more economical if, instead of distributing it in your garden, we put all that material in one place and engineered its disposal so that we could look after it. We are actually going to put it, not in anybody's garden, but in a carefully chosen area. And we are not going to mix it with the top metre of soil; we are going to bury it about 300 m down. The geology will be chosen to prevent the radioactivity from getting out and the waste will be carefully placed in steel and concrete. In addition the waste will eventually decay. So the actual risk to the public will be negligible.

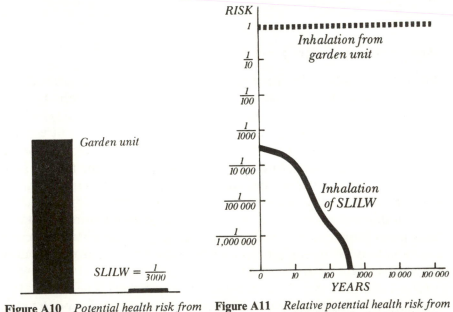

Figure A10 *Potential health risk from inhaling SLILW*

Figure A11 *Relative potential health risk from inhalation of short-lived intermediate level waste*

Long-lived Intermediate Level Waste

Perhaps, then, it's the next category of waste that matters. So let's examine, not the short-lived intermediate level waste that comes mainly from power stations, but the long-lived intermediate level waste that comes mainly from the reprocessing plant.

There are about 4 l of it (about the size of a shoe-box) from each person's lifetime's supply of electricity. The radioactivity is similar in intensity to that from the short-lived intermediate level wastes but with a very important difference: it does not decay away in a few hundred years. It contains very long-lived radioactive elements such as plutonium and americium which are very radiotoxic. In fact it is not dissimilar to the radioactive material in your garden if that were to be separated and concentrated. The uranium and thorium in the garden unit are actually longer-lived than this waste and they are also very radiotoxic.

The waste consists of fuel element cladding, sludges and ion exchange resins and miscellaneous materials contaminated with plutonium. These are currently stored, mainly at Sellafield, but also some at the United Kingdom Atomic Energy Authority's fast reactor establishment at Dounreay, Northern Scotland.

Because they retain their radioactivity for so long it would be inappropriate to bury them near the surface. NIREX therefore plans to bury them, together with the short-lived intermediate level wastes, deep in a stable geological area that would be unaffected by natural phenomena such as erosion that may be caused by ice ages in the distant future. A typical depth would be about 300 m. I have no doubt that when NIREX looks for such a site in a few years' time the cry will again be 'Not In My Back Yard'.

So let us try out the 'thought experiment' on this waste. Since it would be buried deep, let us consider your lifetime's nuclear electricity share of the long-lived intermediate level waste (about 4 l) mixed in the top 300 m of your garden.

This time, the potential risk is not due to inhaling the radioactivity but to the possibility of its getting into drinking water and the food chains. Let us now compare the potential dose and risk from that waste with that from the radioactivity already in the deep garden unit.

The long-lived intermediate level waste would give a potential risk 20 times lower than that from the natural radioactivity there already, and it would reduce in time compared to the potential risk from the deep garden unit (Figures A12 and A13)

Now I can't say that the additional risk (5 per cent) is negligible. It is nonetheless, I think, acceptable especially if you remember we are only doing this in our thoughts. So therefore on long-lived intermediate level

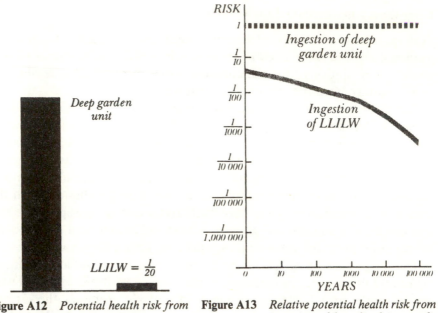

Figure A12 *Potential health risk from ingesting LLILW*

Figure A13 *Relative potential health risk from ingestion of long-lived intermediate level waste*

waste also I must ask the question: 'Why not in my back yard?' or, more precisely, 'Why not in your back yard, provided we bury it in the top 300 m of soil?'

As you see, theoretically we could do so; the extra potential risk is small. But we can do very much better. We do not actually propose to put it in anybody's garden; rather we propose to put it in a carefully chosen area. We are not actually going to mix it in the top 300 m of soil; rather we propose to put it at least 300 m deep. The geology we choose will be ideal for preventing the radioactivity from getting out and the waste will be carefully placed in steel and concrete. The waste will decay slowly with time and we shall still have nature's enormous safety factors protecting us.

Furthermore, owing to the methods we have developed for encapsulating this waste in concrete and encasing it in steel containers we can confidently predict that we shall do better than nature's safety factors. Indeed, until we and the Regulators are quite certain that this is so, we are not even going to bury it at all. I hope, therefore, you will accept that we are being responsible and that there is no need to fear long-lived intermediate level waste either.

High Level Waste

So what about the final category – high level waste? After all, this material – produced during the reprocessing of the spent nuclear fuel – contains some 98% of the radioactivity that is created in making nuclear electricity. So maybe it is this waste that we ought to be worried about.

Let us examine that as the last proposition in this 'thought experiment'. As before, we will start by considering the nature and quantity of this special waste. It is stored as a concentrated liquid at Sellafield (with rather less at Dounreay) but in the next few years it will be vitrified – that is, made into glass. The glass blocks containing the radioactivity will be encased in stainless steel containers. These will be kept in a special air-cooled store to allow much of the radioactivity and the heat emitted from the waste to decay before disposal is considered in at least 50 years' time. In my opinion these wastes will actually be stored for about a century before disposal.

Returning to our 'thought experiment', the quantity of your lifetime's nuclear electricity is, this time, only 0.14 l (about the size of a cigarette packet). Let us imagine that we powder it up and bury it in the top 300 m of ground under your garden. Let us now look at the potential risk from this material in comparison with that from the natural radioactivity which is already there. Now, for the first time, I have a potential risk which is greater than you would get from the deep garden unit of radioactivity (Figures A14 and A15).

It's initially about four times bigger. However, this activity decays in time, and if we store it for about 100 years the potential risk falls below that from the deep garden unit. In fact, it becomes not much different from the long-lived intermediate level waste that we discussed earlier.

But of course we are only doing it in our thoughts to understand the relative potential risk. As with other wastes, we can actually do much better.

We are not proposing to put it in anybody's garden – we are eventually going to put it in a carefully chosen site. We are not actually going to mix it in the top 300 m – we will bury it at least 300 m deep. Ideal geology will be chosen, the waste will be carefully incorporated in glass, encased in metal and surrounded with concrete. We in the nuclear industry propose to store it ourselves for a long time before proposing to bury it anywhere. The special combination of glass, metal and concrete, and the very nature of the radioactive products themselves, make us confident that we have a safety factor much bigger than nature automatically gives us. So the way in which the nuclear industry is proposing to dispose of this and the other nuclear wastes discussed is, in my judgement, safe and ought to be acceptable.

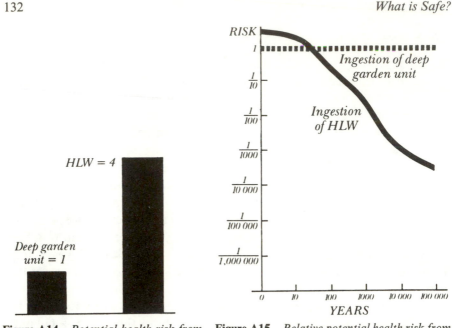

Figure A14 *Potential health risk from ingestion of high level waste*

Figure A15 *Relative potential health risk from ingestion of high level waste*

The Coal Option

I said earlier that I would examine the option of producing your electricity entirely from coal rather than from nuclear power. Now if you elect for that option, I shall first be obliged to charge you more for your electricity, because electricity produced from coal is not as economic as that from nuclear power.

Again in our 'thought experiment', I must ask you to take responsibility for the waste product of burning coal, which is coal ash, and may I remind you that coal ash is naturally radioactive also. Therefore let us do exactly the same comparison of the radioactivity in your garden with coal ash as we have done with nuclear wastes.

A lifetime's supply of electricity produced by coal gives rise to about 20 m³ of ash. Let's suppose that we deliver it to you on your retirement and that it is dug into the top metre of your garden. Then let us compare the potential risk from the radioactivity it contains, with the radioactivity that is already in your garden – the garden unit. (Let's forget about the potential toxic hazard from the chemicals that are also in the ash.)

As a potential risk, the radioactivity in the coal ash is only 10 times less than that in the garden unit and if I plot it with time the radioactivity in the coal ash does not decay either, relative to the garden unit of radioactivity (Figures A16 and A17).

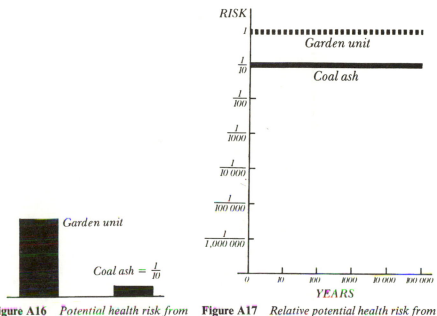

Figure A16 *Potential health risk from inhaling coal ash*

Figure A17 *Relative potential health risk from inhalation of coal ash*

Since the radioactivity in the coal ash is not concentrated, it would be impracticable to put it in concrete or glass. I would just have to dig it into the top metre of your soil. This diagram is a fair representation of the potential risk from the coal ash radioactivity if we actually carried out our 'thought experiment'. I stress that we are talking about potential risks and I remind you that the real risks are considerably less. In fact, the real risks from coal ash are, to all intents and purposes, negligible. That is why we do not treat it as a radioactive waste when we dispose of it or when we make it into building blocks or use it in road construction.

The graph overleaf summarizes all the previous graphs. It shows that the potential risks from the long-lived intermediate level waste and high level waste, put at a suitable depth in your garden, are similar to those from coal ash mixed in the top metre of your garden, and that all the other wastes I have discussed are considerably safer than that.

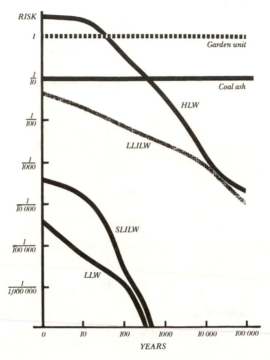

Figure A18   *Summary of the curves. Comparison is of inhalation from shallow material or ingestion of deep material as appropriate*

Further Reading

'Against the Odds', L. Frewer, *Chemistry in Britain*, 1997, January, 21–22.
On the State of the Public Health 1995, Chief Medical Officer, Department of Health, HMSO, London, 1996, ISBN 011-321989-X.
Living with Risk, The British Medical Association, M. Henderson, Penguin, London, 1990, ISBN 0-14-012504-3.
Radiation and Public Perception, Benefits and Risks, eds. J.P. Young and R.S. Yallow, American Chemical Society, Washington DC, 1995, ISBN 0-8412-2932-5.
'Radiation Risk, Risk Perception and Social Constructions', eds. J.B. Reitan, U. Bäverstam and G.M. Kelly, *Radiation Protection Dosimetry*, Nuclear Technology Publishing, Ashford, 1996, **68**(3/4), 155–300, ISBN 1-870965-44-2.
U. Beck, *Risk Society, Towards a New Modernity*, Sage Publications, London, 1992, ISBN 0-8039-8345-X.
Risk, Environment and Modernity, Towards a New Ecology, eds. S. Lash, B. Szerszynski and B. Wynne, Sage Publications, London, 1996, ISBN 0-8039-7937-1.
'Silent but Deadly – Radon', P. Phillips, T. Denman and S. Barker, *Chemistry in Britain*, 1997, January, 35–38.
D. Wilkinson, *Nuclear Waste: Perception and Realities*, Text of Lecture at Imperial College (Chairman of the Radioactive Waste Management Advisory Committee 1978–83), University of London, 5th June 1984.
Understanding Risk. Informing Decisions in a Democratic Society, eds. P.C. Stern and H.V. Fineberg, Committee on Risk Characterization, National Research Council of USA, National Academy Res., Washington DC, 1996, ISBN 0-309-05396.
Living with Radiation, National Radiological Protection Board, HMSO, London, 3rd edn., 1987, ISBN 0-85951-255-X.
Lord Marshall of Goring, *Your Radioactive Garden*, Central Electricity Generating Board, London, 1986.
A.V. Chambers, S.J. Williams, S.J. Wisbey, A.J. Hooper and I.G. Crossland, NIREX Safety Assessment Research Programme, UK Nirex Ltd., Harwell, UK, 1995, S/95/011.
The Way Forward, UK Nirex Ltd., Harwell, UK, 1992.
The Environmental and Ethical Basis of Geological Disposal, Nuclear Energy Agency, Organization for Economic Cooperation and Development, Paris, 1995.
Disposal Facilities for Low and Intermediate-Level Radioactive Wastes: Principles for the Protection of the Human Environment, HMSO, London, 1984, ISBN 011-751775-5.
Biodiversity, The UK Action Plan, UK Department of the Environment, Cm 2428, HMSO, London, 1994.
Diversity of Radioactive Wastes in Deep Repositories, The Royal Society, London, 1994.
P. Saunders, *Managing Radioactive Waste in the European Community*, CEC, Brussels, 1994.

This Common Inheritance, White Paper on the Environment, UK Department of the Environment, Cm 1200, HMSO, London, 1990.

'The Science of Perceiving Risk', I. Taylor, UK Minister for Science, *Chemistry and Industry*, 1996, 2 December, 956.

'Mortality of Professional Chemists in England and Wales, 1965–1989', W.J. Hunter, B.A. Henman and D.M. Bartlett, *Am. J. Ind. Med.*, 1993, **23**, 615–617.

'What are Tolerable Levels of Risk', S. Harbison, *Atom*, 1993, **430**, 37–40.

'Risk Ranking by Perception', E.K. Osei, G.E.A. Amoh and C. Schandorf, *Health Phys.*, 1997, **72**(2), 195–203.

T.R. Lee, *Risk, Perception and Radioactivity, in Public Perception of Radiation*, Association of University Radiation Protection Officers, London, September, 1995, 3–22.

'If you know a better hole.....', A. Blowers, *New Scientist*, 1997, 10 May, 55.

'Cancer Clusters and their Origins', R. Maheswaran and A. Staines, *Chemistry and Industry*, 1997, 7 April, 254–256.

'The Chernobyl Game', J. Surrey, *The Times Higher*, 1997, 30 May, 13.

Population Trends, HMSO, **87**, 1997, ISBN 0-11-620900-3.

The Health of Adult Britain, 1841–1994, HMSO, London, 1997, vols I and II.

'Acceptability of Medical Procedures – Are X-rays Safe Enough?', P.P. Denby, in *Public Perception of Radiation*, Association of Chemistry Radiation Protection Officers, London, September 1995, 61.

'Understanding the Risks', D.J. Ball, *Chemistry and Industry*, 1992, 19 Oct, 776–779.

'Research in European Rock Laboratories', W. Kickmaier and I. McKinley, *Bulletin*, Nagra 1997, 29 April, 29–36.

Risk Reduction – Chemistry and Energy into the 21st Century, ed. M.L. Richardson, Taylor and Francis, Basingstoke, 1996, ISBN 0-7484-0398 1.

'Radon–Cancer Link', G.M. Kendall and C.R. Muirhead (NRPB), *Chemistry in Britain*, 1997, July, 21.

'On the Low Dose Problem in Radiation Protection', K. Becker, response by J. Valentin, and 'The Case of Linearty' by L. Persson, Swedish Radiation Protection Institute Newsletter, 1997, **5**(1), 7–15.

'Hazop and Hazan – Identifying and Assessing Process Industry Hazards', T.A. Kletz, Institution of Chemical Engineers, Rugby, 3rd edn., 1992.

'Risk – The Royal Society Symposium 1997', J. Hinde, *The Times Higher*, 1997, March 14, 18–21.

N.W. Hurst, *Risk Assessment: The Human Dimension*, The Royal Society of Chemistry, Cambridge, 1998, ISBN 0-85404-554-6.

Risk Assessment and Risk Management, eds. R.E. Hester and R.M. Harrison, The Royal Society of Chemistry, Cambridge, 1998, ISBN 0-85404-240-7.

Anthology of Nirex Science Reports, Nirex, Didcot 1994–96, CD-ROM, S/96/101/ROM.

Rethinking Disposal, The Radioactive Waste Management Advisory Committee, HMSO, London, 1998, ISBN 185-112066-1.

Subject Index

Aberfan school disaster, 28
Accidents, 4, 8, 49
  nuclear, 49
  on railway, 61, 44
  on road, 61, 64
  road traffic, 7, 8
  traffic, 49
Acts of God, 68
Acute and chronic doses, 52
Additives, 34
Advisory Committee on the Safety of
  Nuclear Installations
  (ACSNI), 76, 88
Aeroplane flight to Paris, 110
Aflatoxins, 50
AGR reactors, 28
Air travel, 64, 65
Airbags, 111
Airport runways, 46
Alcohol,
  consumption, recommended limits
    of, 47
  intake, 34
  levels in blood, 62
  mis-use, 68
Alcoholic beverages, 41
Aldermaston, 26
Alpha particles, 13, 73
Anaesthetics, 19
Analytical chemistry, 50, 67
Animal experiments, 64
Anodic stripping voltammetry, 84
Anthrax, 82

Arsenic,
  inorganic, 85
  organic, 84
  poisoning, 50
As Low As Reasonably Achievable
  (ALARA), 40, 104
As Low As Reasonably Practicable
  (ALARP), 56, 104
Atlantic Sea, 81
Atom bomb survivors, 17
Atomic Energy Authority (AEA), *see*
  UKAEA
Australia, 16

BANANA, 104
Barium,
  nitrate, 113
  sulfate, 113
BATNEEC, 104
Becquerel, 70
Beer, 23
Belgium, population, 43
Benzpyrene, 82
Beta particles, 13, 73
Bhopal incident, 50
Bicycles, 41
Birth, 4, 19
Blowers, Professor Andrew, 87
Body politic and scientific, 115
Bone marrow, 74
Borehole research, 86
Borosilicate vitrified glass, 80
Botulinus toxins, 50

Botulism, 82
Brain damage, 22
Bristol, 13
British Medical Association (BMA), 2, 35, 66
British Nuclear Fuels plc, 26, 76
Brundtland Commission report, 29, 116
Burns, 12
Butane gas sniffing, 54

Cadmium fumes, 82
Caesium-137, 42, 57, 67
Caffeine, 82
Californian earthquake, 117
Calman, Sir Kenneth, 115, 116
Calories, 19
Cancer, 5, 16, 33, 64
  bladder, 64
  bowel, 16
  breast, 16
  cervical, 16
  prostate, 16
  skin, 3
  stomach, 16
  uterine, 16
Canoeing, 6
Canvey Island, 66
Carbon oxides, 25
  carbon dioxide, 12
  carbon monoxide, 12
Carcinogenesis, 50, 59
Carcinogens, 16
Cardiovascular disease, 5, 16, 19
CATGROIN, 104
CATNIP, 104
Causal link, 5, 62
Central Electricity Generating Board (CEGB), 79
Channel Tunnel, 9
Chemical,
  industry, 35
  speciation, 81, 84, 90, 93
  wastes, 75
Chemicals, 16
  corrosive, 50

flammable, 49
  garden, 52
  kitchen, 52
  toxic, 49
Chernobyl, 12, 27, 28, 38
  deaths, 100
  incident, 42, 66
  plume, 84
  statistical cancer deaths, 67
Chickenpox, 61
Childbirth, 60
Childhood, 6
Chimneys, blocked, 24
Chloride, 113
Cigarette smokers, 98
Civilization, 5
CJD, v
Clean Air Act of 1956, 24
Clothing and footwear, 49
Club activities, 6
'Clusters', v, 27
  of incidents, 114
Coal, 12
  fires, 11
  mining, 24, 49, 68
Coffee, 43
Collective decisions, 30
Colloids, 88
Commercial aviation, 41
Committee on the Medical Aspects of Radiation in the Environment (COMARE), 26
Computer, v
  code verification, 91
  modelling/simulation, 85, 89, 93
Concrete pouring, 24
Congenital abnormalities, 18
Consultants, 15
Contraceptives, 41
Coral snake venom, 82
Cornwall, 13, 98
Cosmic rays, 101
Cost, 8
Cot deaths, *see* Sudden Infant Death Syndrome

Curie, 71
Cycles for elements in nature, 72
Cycling, 7

Darwin, 72
Death, 6, 16
    annual risk of, 49
    cause of, 5, 6
    certificates, 16
    rate, 8, 101
Deaths, 4
    airline passenger, 68
    Chernobyl, 100
    construction industry, 49
    cot, 6
    UK nuclear industry, 64
Decibels, 14
Decommissioning nuclear energy
        plants, 89
Dental amalgam and mercury, 84
Denver, 42
Department of Health, 60, 114
Department of the Environment and
        Transport, 114
Descartes, v
Devon, 13
Diamondback rattler, 82
Diet, 18
Dietary habits, 19
Diethylenetriaminepentaacetic acid
        (DTPA), 96
Diphtheria, 22, 50
Diving, 14
DNA, 38, 58
Dogma, 117
Doll, Sir Richard, 19
Domesday Book, 23
Dose–response relationships, 39, 50,
        52
Double insulation, 11
Dounreay, v, 17, 26, 86
Drigg, 76, 86
Drink, 23
Drinking and driving, 12
Driving, 5
Drowning, 15, 49

Duty of Care, 23, 52
Dysentery, 23

Ear and throat infections, 20
Earthquake, 69, 89
Education, 6
Electric and magnetic fields, 63
Electromagnetic fields, 43
Emitters, 13
Employment, 49
Energy,
    lifetime's supply of, 99
    nuclear, 17
    portfolios, 24
    requirement of average person, 74
    risks from production of, 23
    solar, 68
    and windmills, 110
E-numbers, for food additives, 53
Equity, 40
Estonia, 9, 43
Ethylenediaminetetraacetic acid
        (EDTA), 80, 93
Exercise, 19
Exposure levels, 43

Familiarity principle, 9
Fast-food society, 19
Fear, 1, 5
Film crews, 15
Fire fighting, 41
Fishing, 49
Flixborough incident, 50
Fluoridated water supplies, 30
Flying, 42
Food,
    colouring, 41
    drink and tobacco, 49
    irradiation of, 43, 63
    preservatives, 41
Formaldehyde, 12
Fossil fuels, 24
Fuel rods, 85
Fulvic acids, 94

Gall bladder removal, 19
Gamma-rays, 17, 20, 66, 73

Garden units, 84
  deep, 99
  shallow, 98
Gas,
  mantles, 98
  natural, 24
  rigs, 25
  town, 11
Gastro-intestinal effects of antibiotics,
   61
Geiger counter, 98
General (private) aviation, 41
General practitioners, 15, 21
Generators, electric, 28
Giro cheques (DSS), 68
Gloves, 54
Goggles, 54
Goiânia incident, 42
Gold, 71
Golfer, 14
Gosforth, 87
Gower Peninsular, 13
Gresford colliery disaster, 27
Grimsel, 87
Group activities, 30
Gujarat Dam burst, 27
Gummer, John, 87

Hahn, 71
Hair dyes, 40
Handguns, 41
Hangliding, 64
Harwell, 27
Hazard, 3, 54
Hazard Analysis Critical Control
   Point (HACCP), 60
Health, 46
  of economy, 46
  foods, 15
Health and Safety at Work Act of
   1974, 23, 48
Health and Safety Executive (HSE),
   110, 117
Health Physics Society, 58
Healthcare, 15
Healthy worker syndrome, 49

Heart disease, 64
Heat stroke, 15
Herald of Free Enterprise, 9
Herbicides, 34
Herd immunity, 27
High level wastes (HLW), 75, 76, 77
High-risk factors, 7
High school and college football, 41
Highly radioactive wastes (HLW),
   26
Hinckley Point C nuclear reactor, 57
Hiroshima bomb, 18, 58
Home,
  appliances, 40, 41
  injuries, 10
Homes, 10
Homicide, 61, 64
Hormesis, 52
Horse riding, 15
House of Lords Environment
   Committee, 79
Humic acids, 94
Hunting, 41
Hydrocarbons, 12
Hypocrates, 3
Hypothermia, 12

Ice-age, 89
Indian King cobra, 82
Industrial revolution, 71
Infant mortality, 4
Infection, 4
Influenza, 61, 64
Instant coffee granules, 67
Insurance agreements, 68
Intermediate level wastes (ILW), 75,
   76, 77, 86
International Atomic Energy Agency
   (IAEA), 88
International Commission on
   Radiological Protection
   (ICRP), 18, 39, 54
Intravenous nutrition, 52
Invasive surgery, 16
Iodine-131, 57
Ion plasma analysis, 85

Isotopes, 73

Journalism, 21
Juggernauts, 25

Knill, Sir John, 29, 87

Lamb in Cumbria and Wales, 67
Lanthanide(III), 90
Large construction, 41
Large numbers, 61
Lawyers, ambulance-chasing, 35
Lead-206, 12
Lead content, 111
Lead–acid batteries, 28
Leisure activities, 6
Leukaemia, 61, 64
Life expectancy, 4, 49
    at birth, 6
Lightning strikes, 61, 64
Limestone, 13
Liquid petroleum gas tankers, 25
Long distance running, 15
Lord Marshall, Walter, 79, 97
Low level wastes (LLW), 75, 76, 77
LULU, 104
Luminous dials of watches and clocks,
    98

M25 London orbital motorway, 7
Magnox reactors, 28
Major, John, 112
Masks, 54
Measles, 22, 61
Media,
    manipulation, 31
    reporting bias, 7
Medical diagnostics, 73
Medical radiation, 40
Medical risks, 19
Medical/dental X-rays, 101
Meitner, 71
Mercury vapour, 82
Metal,
    manufacture, 49
    species and colloids, 94
Metals adsorbed on solids, 94

Mexico City liquid petroleum gas
    explosion, 28, 66
Micro-organisms, 66
Middle-classes, 11
Milk, 19
MilliSieverts, 17
Mining, 24
Minister for Science and Technology,
    115
Minister for the Environment (UK),
    102
Minority groups, 46
Modelling,
    computer simulations, 85, 89, 93
    deterministic effects and, 39, 89
    doses and risks, 88
    probabilistic, 89
Moral aspects of nuclear science,
    38
Morals, 47
Morbidity, 5
    of hospital patients, 54
Mortality, 5
Motor vehicles, 41
Motorcycle riding, 41, 42
Mountain climbing, 41
Mountaineering, 6, 15
Multiple safety barrier principle, 76
Mumps, 22
Murders, 64

Nagasaki bomb, 18, 59
National Lottery ticket, 68
National Radiological Protection
    Board (NRPB), 13, 43, 114
Natural disasters, 49
Natural nuclear reactors, 71, 89
Natural waters, 93
Neptunium(V), 90
Nerve gas (Sarin), 40, 82
Neutron bombardment, 71
New England, 42
New York, 42
NIABY, 104
'Not in my back yard' (NIMBY), 30,
    97, 104

NIMTO, 104
NIREX, vi
   Inventory, 84
Nitrilotriacetic acid (NTA), 80, 93
Nitrogen oxides, 12, 25
No adverse Effects Level (NEL), 64
North East Scotland, 13, 98
North Sea oil rigs, 24
Northern Ireland hostilities, 68
Nuclear age, 70
Nuclear Energy Agency (NEA), 29
Nuclear fission, 71, 72
Nuclear fuel cycle, 75
Nuclear Installations Inspectorate
   (NII), 57, 79
Nuclear magnetic resonance imaging,
   32
Nuclear power, 41, 72
   benefits from, 30, 105
   station construction, 25
   and waste, 63, 96, 97
Nuclear radii, 71
Nuclear reactors, 70
Nuclear warheads and tests, 17, 71
Nuclear Waste Disposal Corporation
   (NIREX), 79
Nuclear weapons fallout, 40, 101
Nuts, 67

Obesity, 19, 68
Occupational exposure, 54
Ocean-going fishing, 49
Off-shore oil rigs, 25
Office of Population Censuses and
   Surveys (OPCS), 4
Oil disasters, 28
Optimistic bias, 34
Oral contraception and venous
   thrombosis, 60
Organic chelates, 94
Organic complexes, 94
Organic mercury, 84
Outrage factor, 39
Oxidation states, 95
Ozone, 70, 109
   depletion, 109

Palliatives, 16
Paper-printing industries, 50
Parasites, 66
Parturition, 6
Pascal, 3
Passive smoking, 68
Pedestrian, 7
Peer review, 86, 88
Pennsylvania, 42
Penny coins, 68
Perception, 29, 37
   public, 63
   of radiation risk, 105
   of radwaste disposal, 102
   of risks, 41
Periodic table of elements, 70
Pertussis, 22
Pesticides, 41
Pharmaceuticals, 52
Philosopher's stone, 31
Phosgene, 82
Piles operation, 21
Planet and evolution, 70
Planning Inquiry, 86
Platinum, 71
Plutonium, 73, 90, 93, 95
   in humans, 50
   oxidation states of, 90, 91
   solubility of, 93, 95
   speciation, 83, 90, 91
   toxicity of, 82
Plutonium-239, 93, 95
Pneumoconiosis, 25
Poisoning, 11, 49
   accidental, 64
   arsenic, 50
   strychnine, 50
Police work, 41
Politically active women, 40
Pollution, 16
Polyunsaturated fatty acids, 19
Potassium-40, 98
Power,
   electric (non-nuclear), 41
   hydroelectric, 24
   mowers, 41

Power (*continued*)
  nuclear, 41,72
    benefits from, 30, 105
Power stations,
  coal-fired, 25, 99
  nuclear, construction of, 25
  oil-fired, 110
Pregnancy, 60
Prescription antibiotics, 41
Prey, 19
Professional citizens, 40
Prophylactic agents, 15
Psychometric studies, 40

Quality,
  control, 43
  of life, 5, 30

Radiation,
  background, 17, 67
    levels of, 98
    natural, 64
  damage, threshold for, 58, 63
  doses, 101
  flights to Spain and, 66
  industrial uses of, 73
  perception of risk from, 105
  received working in radiation
      industry, 64
  release of, 61
  shortwave, 19
  in sea waters, 101, 102
  Sun and, 98
  workers, 49
Radioactive half-lives, 73
Radioactive Waste Management
    Advisory Committee
    (RWMAC), 76, 88
Radioactivity,
  Eastern Mediterranean Sea, 102
  Irish Sea, 26, 81, 101, 102
  natural, 101
  Persian Gulf, 102
  Red Sea, 102
Radiological protection, 18
Radionuclide pathways, 72, 73

Radiotherapy, 17
Radium-226, 25
Radon,
  in homes, 12, 40, 63, 109
  information leaflets, 13
Radwaste cross-boundary
    movements, 109
Railroads, 41
Reactor plutonium, 82
Regional variations, 43
Rescue missions, cost of, 33
Respiratory disorders, 12
Risk, 1, 2, 3, 8, 30, 37, 62
  acceptable, 38
  adjustment factors, 37
  avoiders, 54
  benefit and cost, 1
  of death per person, 64
  estimation, 62
  evaluation, 62
  general principles of, 54
  management, 45
  perception of, from radiation,
      105
  research, 36
  science education concerning, 37
  takers, 54
  tolerability; *see also* Tolerability of
      risk (TOR), 3
  tolerable levels of, 56
  vs. cost, 59
  zero, 22, 45, 54
Risk-free society, 38
Risks,
  from chemicals, 50
  collective, 31
  comparative, 61
  drug, 21
  from energy production, 23
  imposed, 33
  medical, 19
  modelling doses and, 88
  negligible, 60
  occupational, 48
  perceived, 41
  personal, 45

Risks (*continued*)
   prioritization of, 46
   recreation, 6, 14
   societal, 45
   transport, 7
   at work, 48
RNA genetic code, 58
Rock Characterization Facility
      (RCF), 86
Rock pores, 90
Rock-climbing, 55
Romans, 23
Routes of entry of contaminants into
      humans, 80
Royal Commission on Environmental
      Pollution, 76
Royal Society of Chemistry, 52
   causes of death survey, 63
Royal Society Report on NIREX's
      proposal, 86
Rubella jabs, 22

Saccharin, 64
Safe, 1, 4, 8, 54, 59, 60
Safe enough, 59
Safety, 45
Safety Degree Unit (SDU), 64
Safety measures, 48
Salt, 19
San Andreas Fault, 117
Savitch, 71
Scandinavian Star Ferry, 9
Seizure, 22
Sellafield, v, 17, 26, 28, 82, 86
Sheep, 67
'Shock-horror-scare' press reporting,
      31, 103, 116
Short-lived intermediate level wastes
      (SLILW), 75
Signal value, 40
Silver, 71
Skiing , 15, 41
Skin, 74
   cancer, 3
   rashes, 21
'Slip, slap, slop', 16

Smogs, London, 3
Smoking, 4, 13, 16, 17, 34, 41, 61, 64
Snow, C.P., 115
Snuff, 3
Social constructions, 38
Social drinking, 33
Sodium-24, 72
Sport, 14
   rugby, 15
   soccer, 15, 64
Sports,
   cars, 14
   dangerous, 54
   fatal accidents in, 49
   spectator, 15
Spray cans, 41
St. Paul's Cathedral, 76
Stairs, 46
Standardized mortality ratio (SMR),
      4, 5
Statistics, 61, 109
   gathering difficulties, 8
Stigma, 40
Stochastic effects, 39
Strassman, 71
Stress, 37
Strikes by lightning, 49
Stripa, 87
Students, 40
Sudden Infant Death Syndrome
      (SIDS), 6
Sugar, 19, 29
Suicides, 11, 64
Sulfur oxides, 12, 25
Sun,
   bathing, 33, 40
   protection, 16
   and radiation, 98
Surgery, 35, 41
Sustainable World, 29
Sweden and radwaste, 86
Sweeteners, 19
Swimming, 14, 41
   pools, 46

Tankers, 28

Taylor, Ian, 115
Television screens, 15
Teratogenesis, 21, 50
Terrorism, 40
Tetanus, 22, 50
Textiles, 49
Theme park visitors, 64
Therapy, 72
Thermal Oxide Reprocessing Plant
    (THORP), 74
Thermodynamic databases, 99
Thiamin, 66
Thorium, 98
Thorium-238, 25
Thorium(IV), 90
Three Mile Island, 27, 42
Threshold Limit Value (TLV), 64
Timber, furniture, 49
Tobacco, 3
    smoke, 12
    smoking, 68
    tax, 17
Tolerability of risk (TOR), 46, 56,
    58
Tono, Japan, 87
Tonsils operation, 21
Trace elements and biochemistry,
    70
Training, 49
Transmission of HIV, 61
Travel, 4
    day-time, 7
    night-time, 7
    Saturday night, 8
    Sunday night, 8
Trust, 43
    increasing and decreasing, 44
    in large institutions, 38
Turin shroud, vi

UK Atomic Energy Authority
    (UKAEA), 29
UK NIREX Ltd., 76
Uranium, 98
    prices, 74
Uranium-238, 12, 25

Uranium(VI), 90
Utopia, 12

Vaccination, 22, 41
    measles, 22
Vaccinations, 61
Validation of models, 93
Variability analysis code, 96
Vegetables, 19
Vehicles, 8
Venetian blinds, 111
Very low level wastes (VLLW), 75
Violence, 4, 61, 64
Visiting a theme park, 49
Visitor centres, 103
Vitamin C, 66
Vitamins, 19
Voltages, 11
Votes, 33

Waste classification, 75
Wastes,
    agricultural, 75
    biological, 75
    chemical, 75
    coal-fired power station, 25, 99
    commercial, 75
    controlled, 75
    difficult, 75
    disposal of, radioactive, 47, 70
    domestic, 75
    hazardous industrial, 68
    high level (HLW), 75, 76, 77
    highly radioactive (HLW), 26
    industrial, 75
    inert, 75
    intermediate level (ILW), 75, 76, 77,
        78
    low level (LLW), 75, 76, 77
    non-controlled, 75
    non-special, 75
    and nuclear power, 63, 96, 97
    pH of, 80
    physical, 75
    radwaste, 75
        disposal of, 47, 70

Wastes, radwaste (*continued*)
    predicted arisings of, 77
    volumes of, 77, 78
  toxic potential of, 112
  very low level (VLLW), 75
  vitrified, 99
Water supplies, 59
Western Culture, 19
Wilkinson, Sir Denys, 81
Williams, D.R., 102
Windmills, 28, 110
Windscale, 74
  pipeline, 81
Wood dressings, 12

Wood-fired stoves, 11, 12
Woodworking, 12
Working in industry, 49
World Commission on Environment
    and Development, 29
World population, 68

X-ray contrasting agent, 114
X-rays, 17, 19, 22, 32, 41, 66
  chest, 42
  medical/dental, 101

Young business persons, 40